Advances in Optofluidics

Advances in Optofluidics

Special Issue Editor

Xuming Zhang

MDPI • Basel • Beijing • Wuhan • Barcelona • Belgrade

MDPI

Special Issue Editor
Xuming Zhang
The Hong Kong Polytechnic
University
China

Editorial Office
MDPI
St. Alban-Anlage 66
Basel, Switzerland

This is a reprint of articles from the Special Issue published online in the open access journal *Micromachines* (ISSN 2072-666X) in 2018 (available at: http://www.mdpi.com/journal/micromachines/special_issues/Advances_in_Optofluidics)

For citation purposes, cite each article independently as indicated on the article page online and as indicated below:

LastName, A.A.; LastName, B.B.; LastName, C.C. Article Title. *Journal Name* **Year**, *Article Number*, Page Range.

ISBN 978-3-03897-095-8 (Pbk)
ISBN 978-3-03897-096-5 (PDF)

Cover image courtesy of Xuming Zhang.

Contents

About the Special Issue Editor

Xuming Zhang is currently an Associate Professor in the Department of Applied Physics, Hong Kong Polytechnic University. He received B.Eng. degree from the University of Science & Technology of China in 1994, M.Eng. degrees from Shanghai Institute of Optics and Fine Mechanics in 1997 and National University of Singapore in 2000, and a Ph.D. degree from Nanyang Technological University in 2006. He has published approximately 100 articles in peer-reviewed journals. His research work has been extensively reported and highlighted by international media and professional magazines. His research interests cover mainly optofluidics, microfluidics, nano-optics, optical sensors, photocatalysis, and artificial photosynthesis.

micromachines

MDPI

Editorial

Editorial for the Special Issue on Advances in Optofluidics

Xuming Zhang

Department of Applied Physics, The Hong Kong Polytechnic University, Hong Kong 999077, China;
apzhang@polyu.edu.hk

Received: 14 June 2018; Accepted: 14 June 2018; Published: 15 June 2018

Optofluidics grew up from the early attempts to integrate optics with microfluidics to exert the benefits of both. Over the last decade, it has made steady progress from the passive interaction of light and liquid media such as liquid waveguides, optical sensors and liquid tunable lenses to the active interaction of photons and liquids such as lasers, particle manipulations and photoreactions. This special issue of Micromachines, entitled *"Advances in Optofluidics"*, collects 9 review articles that update the latest progress of the optofluidics in a broad range of topics, such as droplets, light manipulation, display, refractometry, microcavities and tunable lenses.

One of the core parts of optofluidics is to use liquids as the optical media. Two review articles cover this topic. Yang et al. discussed how to use inhomogeneous liquids to form refractive interfaces for light focusing, or to generate gradient refractive index profiles for advanced effects such as self-imaging, discrete diffraction and optical cloaking [1]. In the other article, Chen et al. narrowed down to a more specific topic—tunable liquid lenses for in-plane light focusing/diverging [2]. They categorized the lenses based on the operation mechanisms and presented their applications in integrated lab-on-a-chip systems for particle trapping and flow cytometry.

Another core part of optofluidics is to use light to measure the change in the liquid media. This special issue has two review articles in this topic. Jian et al. reviewed the recent work on optofluidic refractometry and elaborated different sensing mechanisms/structures and the performance enhancement [3]. In addition, Wang et al. focused on the use of optofluidics to monitor the water quality such as heavy metal, organics, and microbial pollution [4]. This is a new but important area and is worth more exploration.

Droplet microfluidics aims to discretely manipulate tiny volume of fluids. The use of light enables the droplet sensing and manipulation. In this special issue, El Abed reviewed recent developed methods for real-time analysis of droplet size and size distribution, for active merging of microdroplets using light, or for optical probing [5]. Huang et al. reviewed another aspect—passive micromixing in droplets [6], covering the element designs, the analysis methods and the numerical models.

Optofluidic microcavities confine both liquid and light in a tiny space and significantly enhance their interaction, especially the active interaction of photons and liquid media. Along this line, Feng et al. summarized the recent advances in the liquid microlasers and their biochemical sensing applications [7]. This review article categorized the laser structures of optical resonant cavities and classified the active and passive sensors.

Optofluidics is often based on a microchip. In fact, it can also utilize other supporting structures with air channels, for instance, microstructured optical fibers. Shao et al. reviewed the recent progress in this interesting and useful topic and discussed various sensing applications [8].

Optofluidic display has become a hot topic in recently years thanks to the brilliant idea and the bright market prospect. As an expert of this field, Shui et al. elaborated the working principles and device structures of three types of reflective displays, and summarized the optofluidic behavior and the controlling factors [9]. Display is one of the areas that bear the hope of real, impactful

Micromachines **2018**, *9*, 302

application of optofluidics. This review article lays down the technical bases and serve as the guide for other researchers.

Certainly, there are some overlaps among these review articles. For instance, the review of optical sensors may partially cover the optical structures, and the application discussions of microcavities and droplets have to involve optical sensors. In view of the completeness of each individual review article, such an overlap is inevitable. Fortunately, the overlap is minor because each article has its own focal interests, which are different from those of the other articles.

I would like to thank all the authors for their great contributions to this special issue. Sincere appreciation also goes to all the reviewers for their efforts and visions to ensure the quality of review articles.

Conflicts of Interest: The author declares no conflicts of interest.

References

1. Zuo, Y.; Zhu, X.; Shi, Y.; Liang, L.; Yang, Y. Light Manipulation in Inhomogeneous Liquid Flow and Its Application in Biochemical Sensing. *Micromachines* **2018**, *9*, 163.
2. Chen, Q.; Li, T.; Li, Z.; Long, J.; Zhang, X. Optofluidic Tunable Lenses for In-Plane Light Manipulation. *Micromachines* **2018**, *9*, 97. [CrossRef]
3. Li, C.; Bai, G.; Zhang, Y.; Zhang, M.; Jian, A. Optofluidics Refractometers. *Micromachines* **2018**, *9*, 136. [CrossRef]
4. Wang, N.; Dai, T.; Lei, L. Optofluidic Technology for Water Quality Monitoring. *Micromachines* **2018**, *9*, 158. [CrossRef]
5. Hayat, Z.; El Abed, A.I. High-Throughput Optofluidic Acquisition of Microdroplets in Microfluidic Systems. *Micromachines* **2018**, *9*, 183. [CrossRef]
6. Chen, C.; Zhao, Y.; Wang, J.; Zhu, P.; Tian, Y.; Xu, M.; Wang, L.; Huang, X. Passive Mixing inside Microdroplets. *Micromachines* **2018**, *9*, 160. [CrossRef]
7. Feng, Z.; Bai, L. Advances of Optofluidic Microcavities for Microlasers and Biosensors. *Micromachines* **2018**, *9*, 122. [CrossRef]
8. Shao, L.; Liu, Z.; Hu, J.; Gunawardena, D.; Tam, H.-Y. Optofluidics in Microstructured Optical Fibers. *Micromachines* **2018**, *9*, 145. [CrossRef]
9. Jin, M.; Shen, S.; Yi, Z.; Zhou, G.; Shui, L. Optofluid-Based Reflective Displays. *Micromachines* **2018**, *9*, 159. [CrossRef]

micromachines

MDPI

Review

Light Manipulation in Inhomogeneous Liquid Flow and Its Application in Biochemical Sensing

Yunfeng Zuo, Xiaoqiang Zhu, Yang Shi, Li Liang and Yi Yang *

School of Physics and Technology, Wuhan University, Wuhan 430070, China; zuoyf@whu.edu.cn (Y.Z.); daveria@whu.edu.cn (X.Z.); shiyang@whu.edu.cn (Y.S.); lianglill@whu.edu.cn (L.L.)
* Correspondence: yangyiys@whu.edu.cn; Tel.: +86-027-6875-2989 (ext. 8103)

Received: 29 January 2018; Accepted: 26 March 2018; Published: 2 April 2018

Abstract: Light manipulation has always been the fundamental subject in the field of optics since centuries ago. Traditional optical devices are usually designed using glasses and other materials, such as semiconductors and metals. Optofluidics is the combination of microfluidics and optics, which brings a host of new advantages to conventional solid systems. The capabilities of light manipulation and biochemical sensing are inherent alongside the emergence of optofluidics. This new research area promotes advancements in optics, biology, and chemistry. The development of fast, accurate, low-cost, and small-sized biochemical micro-sensors is an urgent demand for real-time monitoring. However, the fluid flow in the on-chip sensor is usually non-uniformed, which is a new and emerging challenge for the accuracy of optical detection. It is significant to reveal the principle of light propagation in an inhomogeneous liquid flow and the interaction between biochemical samples and light in flowing liquids. In this review, we summarize the current state of optofluidic lab-on-a-chip techniques from the perspective of light modulation by the unique dynamic properties of fluid in heterogeneous media, such as diffusion, heat transfer, and centrifugation etc. Furthermore, this review introduces several novel photonic phenomena in an inhomogeneous liquid flow and demonstrates their application in biochemical sensing.

Keywords: optofluidics; inhomogeneous medium; light manipulation; biochemical sensing

---

1. Introduction

The term 'optofluidics' appeared in 2003 at the California Institute of Technology in Pasadena, owing to the development of microfluidics [1]. Microfluidics is a promising field aimed at fluidic manipulation with various applications in fields such as biochemistry technology, molecular analysis, chemical synthesis, and energy conversion [2–7], and especially in field of single-cell biology [8–10]. Pagliara et al., developed a microfluidic assay to confine single cells. This design provided an efficient tool for single cell detection [11]. Optofluidics is a new analytical field that focuses on the integration of optics and microfluidics, providing many unique characteristics for enhancing the performance and simplifying the design of micro-electromechanical systems [12–14]. Over the past decades, this new field has rapidly developed and has been applied in many areas, such as biosensors, biomedical analyses, energy production, optical imaging and many other optical systems [15–22]. The manipulation of light, such as in light routing, focusing, bending etc. plays an important role in lab-on-a-chip optofluidic systems. Recent advancement in optofluidics has demonstrated a new class of microsystems that exploits microfluidic flow to manipulate light in the microchannel, realizing various optical devices and functionalities [23] such as liquid microlenses [24,25], prisms [26,27], gratings [28,29], switch [30,31], and dye lasers [32,33].

For the purpose of light manipulation, optofluidic lab-on-a-chip techniques have several novel merits that cannot be found in conventional solid-based optical systems [1,12,13]. Optofluidics makes full use of liquids to manipulate light beams. Liquids are natural optical materials that possess greater tunability in their refractive index and a greater flexibility in shape than their solid equivalents. Pure liquid optofluidic devices fabricated by polydimethylsiloxane (PDMS) have several useful characteristics. First, the optical property of the fluid media and refractive index (RI) distribution can be easily changed by replacing one fluid with another or changing the flow rates. Second, the interface between the two fluids can be optically smooth and that can reduce the propagation loss of the light beams. Finally, there is the ability to create a gradient refractive index (GRIN) through the process of liquids diffusion. The optical properties, such as refractive index, absorption, and fluorescence etc., and the physical properties, such as the magnetic susceptibility and electrical conductivity of the optofluidic devices can be changed easily, dynamically, and continuously. These properties can be used to design novel devices.

As mentioned above, optofluidic devices and systems can be tuned by changing the refractive index distribution of its liquid. Conventional light manipulation techniques change the refractive index by introducing an external electrical field, magnetic field, temperature field, acoustic field, or mechanical strain. Optofluidic approaches such as pumping and mixing can be used for changing the liquids or its composition. If the liquid is a solution, by integrating a concentration generator and mixer in the device, the composition and concentration can be adjusted readily (e.g., the refractive index of $CaCl_2$ solution can be changed from 1.334 to 1.445). In principle, any liquids that are compatible with the microchannel and process excellent light stability can be used to form the optofluidic devices. The refractive index of the common liquids ranges from 1.300 (2,2,2-Trifluoroethanol) to 1.749 (methylene iodide). Since the flow streams of the optofluidic device can be replaced by pumping different liquids, a large relative refractive index change σ ($\Delta n/n$) can be achieved. Where σ is defined as the ratio between the RI difference and the RI of the initial liquid (e.g., if replacing deionized water by Benzothiazole, $\sigma = 0.23$). In an extreme case, replacing air by a liquid, a relative high $\sigma \sim 1$ can be achieved, while other light manipulation techniques suffer from a small value of σ owing to the fixed optical material. Table 1 demonstrates the performance of the different light manipulation techniques [1].

Table 1. Comparison of optofludics with other light manipulation techniques in relative refractive index change (*D*) and responding time (τ) [1].

Technology	σ ($\Delta n/n$)	$\emptyset$(s)
Optofluidics	1	10^{-3}
Liquid crystal	10^{-1}	10^{-3}
Injection current	10^{-2}	10^{-9}
Temperature	10^{-2}	1
Photorefractive	10^{-3}	10^{-1}–10^{-5}
Electro-optic (10 kV/cm)	10^{-3}	10^{-12}
Photoelastic/Acousto-optic (10 W)	10^{-4}	10^{-6}–10^{-7}

In this review, we provide an overview of optofluidic lab-on-a-chip techniques based on an inhomogeneous liquid flow for light manipulation. First, we describe the fundamental concepts and principles associated with liquids and light manipulation in optofluidic systems. Then, previous optofluidic lab-on-a-chip techniques will be categorized in two kinds from the perspective of manipulation of liquid-liquid interface and manipulation of the liquid gradient refractive index distribution. In pure liquid optofluidic systems, the geometric interface between liquids and the refractive index distribution of liquids dominate the light manipulation. Finally, we introduce several novel photonic phenomena due to the interaction of fluids and light in micro- or nano-scale and demonstrate their application in micro-/nano-optical devices and biological/chemical microsensors, etc. We then discuss the outlook, research trends in light manipulation in an inhomogeneous liquid flow and its potential applications in new and ground-breaking research areas.

2. Fundamental Concepts and Principles

As stated above, optofluidics is the combination of microfluidics and optics. The fundamental goal of light manipulation in optofluidics is to form a proper refractive index profile through manipulating the fluid dynamics processes in a microchannel. For the purpose of light manipulation in an inhomogeneous liquid flow, precise control of the liquid is of vital importance. The physical phenomena that exist in microfluidic systems also occur in optofluidic systems. To reveal the fluid dynamics processes in optofluidic devices and differentiate primary from secondary, several dimensionless numbers are used for expressing the ratio of the fluid dynamics phenomena [34].

2.1. The Reynolds Number

To research the flow streams in a microchannel, the Reynolds number (*Re*) is frequently involved to characterize different flow types. It is one of the most fundamental concepts in microfluidics.

$$Re = \rho V L / \mu \tag{1}$$

where ρ is the fluid density, V is the average flow velocity, L is the channel hydraulic diameter and μ is the kinematic viscosity of the fluid. Hydraulic diameter is defined as $L = 4A/P$, where A is the cross-sectional area of the channel, P is the wetted perimeter of the cross-section. When $Re < 2300$, the state of the flow is laminar and is generally smooth and predictable. As *Re* increases, the first features of inertia become apparent. When $Re > 4000$, the flow becomes an unpredictable, irregular turbulent flow [35]. At the microchannel scale, *Re* is so small that the flow is laminar, which provides a stable fluidic condition for optical waveguide.

2.2. The Dean Number

In a condition of relative high *Re*, the inertia role of the fluids cannot be ignored. When the fluid is moving through a curved microchannel, whose radius of curvature *R* is comparatively much larger than the hydraulic diameter of the microchannel *L*, centrifugal forces on the liquid flow will give rise to a secondary motion. In the situation of the secondary Dean flow, the fluid in the center of the microchannel will be driven to the outer side of the cured microchannel. While the fluid close to the channel wall will be swept towards the inside [36]. The Dean flow can be described and defined by a dimensionless Dean number (*De*), which can be expressed as:

$$De = \delta^{\frac{1}{2}} Re \tag{2}$$

where $\delta = L/R$ is a geometrical parameter. When the liquids are designate, the feature size of the curved microchannel and the *Pe* of the liquid streams determine the Dean number and the liquid transverse movement. *De* reflects the relative relation between centrifugal forces and viscous forces.

2.3. The Convective–Diffusive Transport

The inhomogeneous liquid flows in optofluidic devices are prominently governed by the diffusion and convection process. And the convection–diffusion equation is applied to describe the convective–diffusive transport [37]. This equation is a combination of the diffusion and convection equations, and describes the transfer of molecules, energy, or other physical quantities inside a physical system.

The general equation can be formulated as:

$$\frac{\partial C}{\partial t} = D\nabla^2 C - U\nabla C + S \tag{3}$$

where C is the concentration of the liquid, t is time, D is the diffusion coefficient of the solute, and U represents the average velocity in the microchannel. S describes the contribution of the chemical reaction. In almost every situation of optofluidics, there is no chemical reaction between liquids, and it has $R = 0$. The first term on the right side represents the diffusion process while the second term corresponds to the convection process. For a steady-state and passive flow ($S = 0$), the concentration does not change with time, thus the term $\partial C/\partial t$ leads to zero. In most optofluidic devices, the control of convective–diffusive transport is the fundamental principle and the main method of light manipulation. The convection process plays an important role in optofluidic devices based on the manipulation of the liquid-liquid interface. While in the gradient refractive index (GRIN) and the optofluidic transformation optical devices, both of which are based on manipulation of the gradient refractive index, the diffusion process is dominant.

2.4. The Péclet Number

The relative importance of convection to diffusion is characterized by the dimensionless Péclet number:

$$Pe = UW/D \tag{4}$$

where U is the average velocity, W is the width of the microchannel and D is the diffusion coefficient. The competition between convection and diffusion, embodied in the Péclet number, forms the basis of a number of optofluidic techniques for light manipulation.

3. Manipulation of the Liquid-Liquid Interface

A great number of optofluidic devices rely on the interface formed between streams of flowing liquids in a microfluidic channel. The inhomogeneous liquid flow and the step-refractive index distribution are formed by manipulating the liquid-liquid interface through various approaches such as hydrodynamic focusing, Dean flow, and two-phase flow etc. [38].

3.1. Hydrodynamic Focusing

In order to achieve precise control of the liquid-core/liquid-cladding system, the concept of hydrodynamic focusing is introduced. Various optofluidic devices are designed based on hydrodynamic focusing, such as step-refractive index waveguides, prisms, lens, and optical switch etc. A typical hydrodynamic focusing model is generally made of three flow streams: a core flow and two cladding flows [39]. In a simplified model, the micro-channel and velocity profile are symmetrical. The two laminar sheath flows with equal flow rates focus the core flow stream, the width of which can be adjusted from a few micrometers to hundreds of micrometers. Under a low Reynolds number the ratio between the width of the focused (W_f) flow and that of the main channel (W_m) is expressed as:

$$\frac{W_f}{W_m} = \frac{Q_i}{\alpha(Q_i + Q_{c1} + Q_{c2})} \tag{5}$$

where Q_i and $Q_c(Q_{c1}, Q_{c2})$ are the flow rates of the core and the sheath flow streams, respectively. α represents the velocity ratio $\alpha = \overline{v}_f/\overline{v}_0$. The equation reveals that the widths of the focused streams depend on the ratio of Q_{core} and Q_{clad} in the microfluidic channel under the effect of hydrodynamic force. By taking the convection–diffusion transport equation into account and applying it in the hydrodynamic focusing model, the normalized concentration distribution for the core flow and the cladding flows under dynamic equilibrium-state in the microchannel can be expressed as [40]:

$$C_{core}^*(x^*, y^*) = r + \frac{2}{\pi} \sum_{n=1}^{\infty} \frac{\sin(nr\pi)}{n} \cos[y^* n\pi] \times \exp\left[\frac{1}{2}x^*\left(Pe - x^*\sqrt{Pe^2 + 4n^2\pi^2}\right)\right] \tag{6}$$

$$C_{clad}^*(x^*, y^*) = 1 - r + \frac{2}{\pi} \sum_{n=1}^{\infty} \frac{\sin[n(1-r)\pi]}{n} \cos[(1 - y^*)n\pi] \times \exp\left[\frac{1}{2}x^*\left(Pe - x^*\sqrt{Pe^2 + 4n^2\pi^2}\right)\right] \tag{7}$$

where Pe is the Péclet number and r is the initial interface boundary ratio between the core and cladding fluids $r = 1/(1 + 2\beta\kappa)$ ($\beta = \mu_1/\mu_2$), the dynamic viscosity ratio between the cladding and the core flows, and $\kappa = Q_2/Q_1$, the ratio of the flow rates of the cladding and the core streams). As the concentration of the solute is demonstrated, the index profile $n(x^*, y^*)$ can be formulated as:

$$n(x^*, y^*) = [C^*_{clad}(x^*, y^*) \times n_{clad}] + [C^*_{core}(x^*, y^*) \times n_{core}] \qquad (8)$$

In a condition of relative high *Pe*, the convection process dominates the fluid transportation in the microchannel. The RI distribution appears in a state of step-index profile. By changing the flow rates of the three independent flow streams, the width of the focused core flow can be adjusted, ranging from hundreds of micrometers to a few micrometers, even down to 50 nanometers [41]. The position of the focused core flow can be adjusted by introducing cladding flows with different flow rates. Based on hydrodynamic focusing, various optical devices and elements have been realized, such as waveguides and lenses [42–45]. As shown in Figure 1a Wolfe et al., demonstrated a design of pure liquid optical waveguides [42]. These waveguides were constructed by using a three flow streams system, which is a typical hydrodynamic focusing module. An aqueous solution 5M $CaCl_2$ ($n = 1.445$) and deionizer water ($n = 1.333$) were introduced as the high RI core flow and the low RI cladding flows, respectively. Through adjusting the fluid flow rate, the width and the position of the core flow can be reconfigured to allow the flexible manipulation of light dynamically and continuously in real-time. Manipulating the rate of flow or changing the ingredient of the liquids will modulate the optical and physical properties of the optofluidic systems.

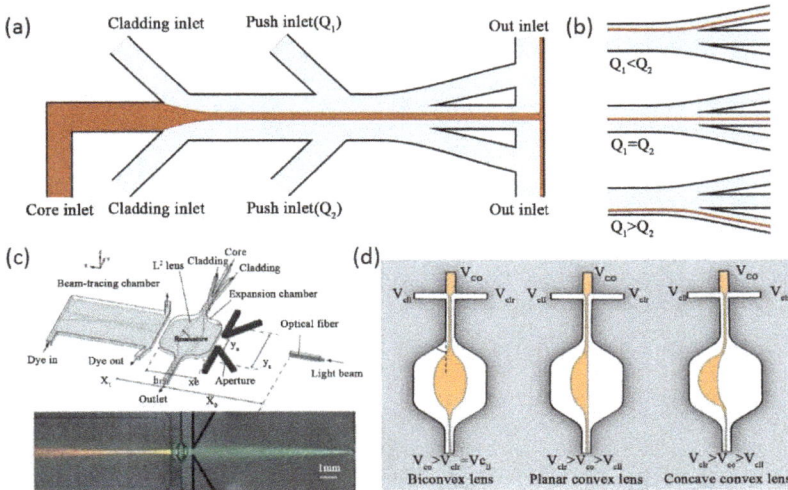

Figure 1. (**a**) Schematic of the liquid-core/liquid-cladding waveguide; (**b**) Verification of light output switching by changing the flow rates; (**c**) Schematic of a pure liquid lens formed by hydrodynamic focusing of three flow streams in a specially designed chip; (**d**) Schematics of the formations of the tunable liquid microlenses. Images reproduced with permission from [43,44].

By changing the flow rates of the cladding flows, the path of the core flow and then that of the light can be determined. An extra set of inlets were added to this liquid waveguide downstream of the initial inlets. These extra inlets were independently controlled to switch the core flow from one output to another without deforming the core flow at the junction point. Thus, this waveguide will maintain a low level of optical loss during output switching. The three switch states were shown in Figure 1b. The output light was controlled easily by changing the flow rates of the "push" inlets. The responding time of this pure liquid switchable waveguide was about 2 s.

As shown in Figure 1c, Tang et al., first demonstrated a pure liquid lens formed by the hydrodynamic focusing of three flow streams in a specially designed chip [43]. The liquid lens was formed in an expanded chamber after the hydrodynamic focusing occurred. The liquid-liquid interface of the core flow and the cladding flows fill the shape of the expanded chamber. In order to maintain the stability and the ideal lens shape, the length-to-width ratio of the expanded chamber was chosen to be 1:1. These lenses process high tunability as different lens shape and the curvature radius can be realized through adjusting the three flows. In the condition of a fixed total flow rate and equal flow rates of both cladding flow liquids, the curvature radius of the convex lens increased alongside an increase in the flow rate of the core flow. Thus, a wide range of focal distance of the liquid lens from ~12 mm to ~6 mm can be obtained through the variation of the rate of the core flow. Seow et al., also reported a similar design for light collimation and focusing [44]. Three types of liquid lens have been demonstrated by tuning these three flow streams as shown in Figure 1d. $CaCl_2$ solution was used as the high RI core liquid ($RI = 1.46$) and the flow rate was V_{core}. Isopropanol solution ($RI = 1.33$) was used as the low RI cladding liquid with flow rates of V_{cl1} and V_{cl2}, respectively. If the flow rates of these two cladding flow streams are the same ($V_{core} > V_{cl1} = V_{cl2}$), a liquid biconvex lens will be formed in the expanded chamber. When V_{cl2} was increased and higher than V_{core}, the curvature radius of the right interface increased at the same time. A planar convex lens will be formed when the curvature radius reaches infinity. By further increasing the flow rates of the right inlet V_{cl2}, it will be reconfigured as a concave convex lens.

Although 2D hydrodynamic focusing is suitable for many applications, these 2D optofluidic devices suffer from large optical loss in vertical direction. The liquid distribution of the 2D devices can only be tuned in the horizontal direction. Considering a microchannel with a width of 200 μm and a height of 100 μm, the width of the core flow can be adjusted from ~200 μm to a few micrometers, however, the height of the core remains 100 μm. The interaction between the microchannel and the core flow limits the performance and tunability of the optofluidic systems. Recent advancements in the fabrication of complex microchannels make it possible to fabricate 3D optofludic devices. A curved liquid channel can also be used for forming 3D liquid systems through the effect of Dean flow.

3.2. Dean Flow

The phenomenon of Dean flow is a novel tool for researchers, which makes it possible to control liquid in the vertical direction. Various novel optofluidic devices have been reported based on Dean flow, such as 3D drifting waveguides [46,47], cylindrical microlens in the Z-axis [48], 3D dye laser [49], and 3D lens [50] etc. Compared with traditional 3D structures dependent on the intricate fabrication and design of microchannels, optofluidic devices based on Dean flow poses the advantages of easy fabrication and high tunablility. Taking the advantages of Dean flow, Mao et al. demonstrated a cylindrical microlens [47]. The liquid cylindrical microlens was generated by the interface between a 5 M $CaCl_2$ solution ($RI = 1.445$) and deionized water ($RI = 1.333$). A curved liquid interface was formed through the centrifugal force generated by the Dean flow. When the two liquid flows moved through the 90 degree curved microchannel, the inner flow of the $CaCl_2$ solution bowed into the DI water and a secondary motion arose in the cross-section of the microchannel. The curved interface between the two liquids formed a planar convex cylindrical lens, as shown in Figure 2a. By changing the flow rates of each flow stream, the geometry of the liquid-liquid interface can be easily modulated, which means that the optical properties of the liquid lens are governed by the flow rate. Figure 2b shows the microscopy images and 3D intensity plots of the focused light spots at different flow rates, i.e., 0 μL/min, 150 μL/min, 250 μL/min. This cylindrical lens based on Dean flow is still a 2D optical device.

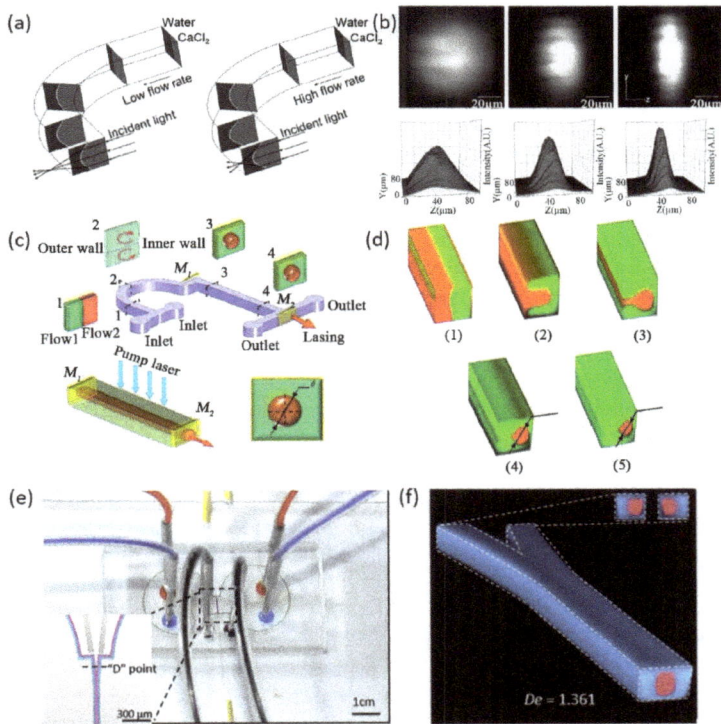

Figure 2. (**a,b**) The pure liquid optofluidic lens. The mechanism of the hydrodynamically tunable optofluidic cylindrical microlens through Dean flow; (**c,d**) The schematic illustration of the 3D liquid waveguide dye laser; (**e,f**) Switchable 3D optofluidic Y-branch waveguides tuned by Dean flows. Images reproduced with permission from [47,49,51].

Through the effect of Dean flow and proper channel design, a series of 3D liquid waveguides and lenses have been demonstrated. Considering a 3D liquid waveguide or lens, the core flow stream was wrapped by the cladding flows. The transverse flow motion generated in the curved microchannel is the main cause for coating the core flow. There are two main factors that count for the realization of the 3D liquid core-liquid cladding structure. First, a relatively high Dean number is necessary to provide a large inertial centrifugal force. From the definition of the Dean number, it is feasible to increase the De by increasing the width of the channel (w) or decreasing the radius of curvature of the curved channel (R) or increasing the Reynolds number (Re) of the flow streams. Second, a long enough flow path is also an essential condition, which provides enough time for the transverse movement of the Dean flow under the effect of the centrifugal force. Another design recently reported by Yang et al., is a tunable dye laser based on a 3D optofluidic waveguide formed by centrifugal Dean flow, as shown in Figure 2c,d. In this optofluidic dye laser, a 3D liquid waveguide formed by two Dean flows acts as the gain medium. The core flow of the 3D waveguide was filled with laser dye. To fulfill the two main factors for the realization of the 3D liquid waveguide, the curved microchannel for Dean flow was designed with a substantially large curvature of radius (R = 2 mm) across 180 degrees. Once the microchannel was fabricated, the flow rates of the inlet flow streams would determine the formation of the 3D liquid waveguide. If the flow rates are relatively low, the inertia role of the fluids will not make a big difference. The inner flow is not completely wrapped by the outer flow. If the flow rates are too high, the position of the two flows will exchange. A proper 3D waveguide will not generate. The effect of centrifugal forces in a microchannel at different flow rates was shown in Figure 2d. Compared with traditional 2D optofluidic dye lasers, this 3D tunable dye laser possessed a higher energy output and a lower lasing threshold. The output energy of this 3D optofluidic dye laser can be varied by changing the flow rates of the two Dean flows in real-time. This liquid laser has the potential to provide a versatile tool for biosensors in optofluidic systems. Li et al., also designed a switchable 3D optofluidic Y-branch waveguide via Dean flows [51], as shown in Figure 2e. Two symmetrical curved microchannels with multi-curvature of radius, small ($R = 1.5$ mm, *arc* = $180°$) and large ($R = 2.5$ mm, *ace* = $265°$), and one Y-junction can be designed. Two independent 3D liquid waveguides can be formed through the separate curved microchannels. When coming across at the Y-junction, these two individual 3D liquid waveguides will combine into one main waveguide, Figure 2f. This Y-branch waveguide can guide light and realize switching the input light by adjusting the state of the independent 3D liquid waveguides, which are under the control of the flow rates of each side. The optical and transmission characteristics of the main 3D liquid waveguide are identical with multimode fiber. This device possesses large tunability and reconfigurability. The relative intensity of the output light at each branch can be adjusted from 1 to 0. When transmitting light with a wavelength of 532 nm in a situation of a 1:1 output ratio, the transmission loss of this 3D optofluidic Y-branch waveguide was estimated to be 0.97 dB. And the light spots at the output remained at a low level of deformation.

Aside from 3D liquid waveguides, Dean flow also allows for the design of 3D liquid lens. Rosenauer et al., presented the first tunable liquid biconic lens with the ability to focus light three-dimensionally [50], shown in Figure 3a. The light focusing in the direction of the X-axis was realized through the fabrication of an expanded microchannel, which shared the same method as in Figure 1c,d. In the direction of the Z-axis, the liquid interfaces of the 3D biconic lens were generated by inserting two curves (90 degrees). The combination of the vertical lens based on Dean flow and the horizontal lens based on the asymmetrical expanded chamber provides a high level of lens tunability. The design with separating functional parts also allows non-correlational modulation of the lateral and transversal curvatures of the lens. An ingenious design of a three-dimensional liquid biconvex lens combined by two symmetrical curved microchannels and an expanded circle chamber was demonstrated by Liang et al., to detect the living cells in flow streams [52], as shown in Figure 3b. Through the auxiliary curved microchannels, a structure of 3D-focusing of the core flow was formed based on Dean flow. As the 3D focused core stream flows into the expanded chamber, it widened in the horizontal and vertical directions and became biconvex in shape. Figure 3c shows the formation of the 3D lens under different flow rates. A wide range of variable focal lengths from 3554 μm to 3989 μm was achieved by changing the flow rates. The 3D liquid lens possesses a large numerical aperture ranging from 0.175 to 0.198, thus providing a higher imaging definition over a traditional objective lens. The images of two living cells, sp2/0 and NB4, were captured through the 3D liquid lens. This kind of 3D biconvex lens has the potential for application of real-time cell imaging and analysis in optofluidic systems.

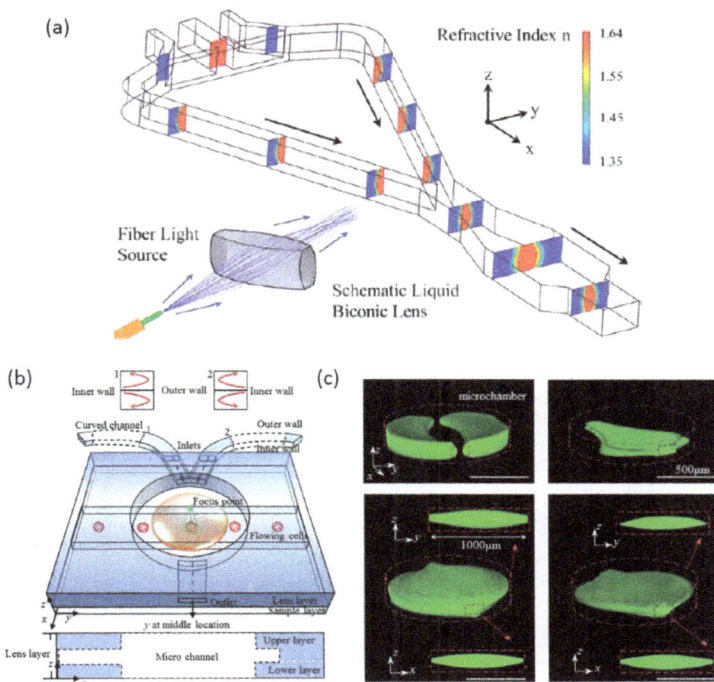

Figure 3. Schematic of the 3D fluidic lens (**a**); (**b**) Schematic diagram of the switchable 3D liquid–liquid lens for cell images; (**c**) The formation of the 3D lens under different flow rates. Images reproduced with permission from [50,52].

4. Manipulation of the Gradient Refractive Index

An inhomogeneous liquid flow in optofluidic devices can be realized not only by controlling the liquid-liquid interface, but also by manipulating the gradient refractive index (GRIN). The GRIN in a microchannel can manipulate light propagating both perpendicular and parallel to the flow direction. Alternative approaches of creating GRIN in optofluidic devices are to form a concentration gradient or a temperature gradient.

4.1. Liquid Diffusion and Transformation Optics

Miscible liquids and their interdiffusion can be of significant use in designing optofluidic devices. The diffusion process of two liquids is a unique characteristic that cannot be found in solid-based devices [53,54]. More concretely, liquid diffusion can create a concentration gradient, thus, forming a refractive index gradient. Here, diffusion becomes an advantage instead of a drawback. The distribution of the GRIN profile in the fluidic systems can be adjusted trough changing the flow parameters, such as *Pe*, or replacing different types of liquids that possess different refractive index and diffusion coefficients. The modulation of GRIN provides great flexibility and tunability to light manipulation for optolfuidic systems. For example, an optical splitter based on the merging of two parallel liquid waveguides has been demonstrated [53]. The diffusion, the refractive index, and the coupling degree between two separated waveguides were determined by the rates of flows. When the flow rate is slow enough to allow full diffusion of the two core flow streams in a microchannel, the two parallel liquid waveguides will smoothly merge into a single waveguide. The input beam will be split into two output beams with an intensity ratio of 1:1 when light propagates in the opposite direction of the flow streams. Unlike a conventional beam splitter, the split ratio of the optofluidic beams splitter can be dynamically tuned. Changing the flow rate changes the gradient of the refractive index, and thus the output ratio and intensity of the liquid beam splitter.

For a typical straight liquid waveguide consisting of three laminar flows, the refractive index profile in a microchannel can be illustrated by Equation (8). Through the manipulation of flow streams, a GRIN distribution can be formed to guide the flow of light. Mao et al., reported a tunable liquid gradient refractive index optofluidic lens for light focusing [54]. Instead of using the step-index interface between curved fluids, a GRIN across the liquid medium was applied to focus and bend light beams, as shown Figure 4a. A hyperbolic secant RI profile, which was suitable for light focusing, was established through the diffusion of $CaCl_2$ solute between the sandwiched core flow ($CaCl_2$ solution) and cladding flows (DI water). Both the focal length and the output direction of light can be tuned by changing the flow rates, the two corresponding working states were shown in Figure 4b. This design of GRIN microlenses has two degrees of freedom, which provides large flexibility and novel functionality for light manipulation. In a later design, Liu et al., reports an optofluidic lens with low spherical and low field curvature aberrations, as shown in Figure 4c [55]. A hyperbolic secant (HS) refractive index profile was generated by adjusting the diffusion between ethylene glycol and deionized water. The spherical aberration in this optofluidic lens HS profile is much lower than that in the former design of liquid GRIN lenses. Owing to the small spherical aberration, the optofluidic lenses have found applications in the manipulation of light source array and multiplexed detection, as shown in Figure 4d.

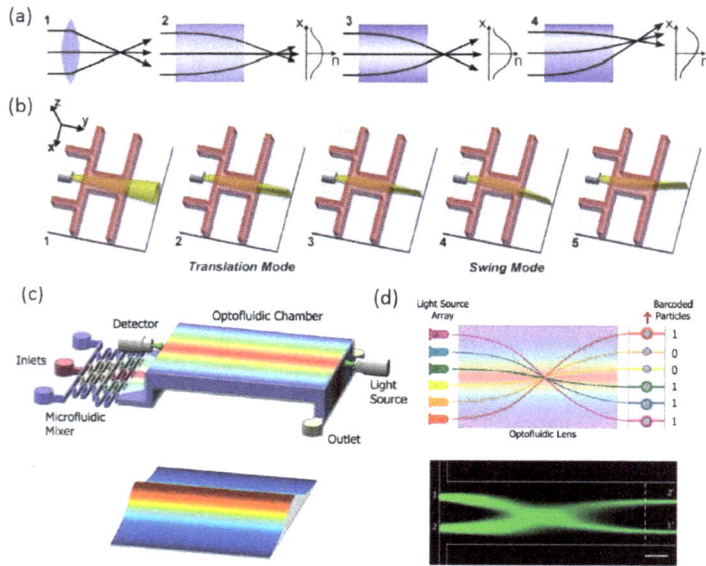

Figure 4. Principle and design of the liquid gradient refractive index L-GRIN lens (**a**,**b**); (**c**) Schematic illustration of the optofluidic lens with low spherical and low field curvature aberrations and its Stable RI distribution; (**d**) Schematic illustration of the potential application in multiplexed detection. Images reproduced with permission from [54,55].

Aside from light focusing and bending, the GRIN of liquid flows also allows for light interference and diffraction. Shi et al., reported a tunable multimode interference (MMI) device as shown in Figure 5a,b [56]. The MMI device consists of two main modules: a GRIN liquid lens and a step-index liquid/solid interface. Three liquid flows with relatively low flow rates were injected to form a gradient RI, thus focusing light and realizing the modulation of MMI. Owing to the chosen low flow rate, full diffusion mixing was realized in the first part of the hybrid waveguide. In the latter part of the optofluidic waveguide, the liquid solution became homogenous so as to form a step-index distribution with the wall of the microchannel. The RI of the core and cladding flows were chosen at 1.432 and 1.410, which were both higher than that of the microchannel (1.405), to obtain a relatively low optical loss of the liquid system. The refractive index profile suited for MMI of the hybrid waveguide was shown in Figure 5a. Figure 5b shows the light focusing and interference patterns at different positions, which was in good agreement with the simulation result. In addition, the period of MMI can be modulated by simply adjusting the flow rates or RI.

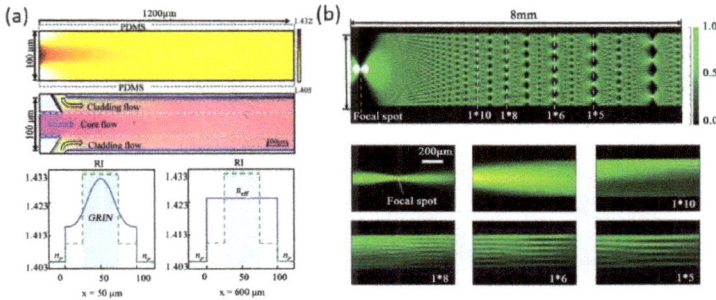

Figure 5. Self-imaging effect by MMI in the hybrid optofluidic waveguide. (**a**) Stable RI distribution in the main channel; (**b**) The self-imaging interference pattern. Images reproduced with permission from [56].

Transformation optics is a fantastic kind of mathematical technique that provides a means to design complex artificial media using the invariance of Maxwell equations in certain coordinate transformations [57]. The artificial media with spatially changing permeability and permittivity offers precious control of the flow of electromagnetic waves. The most striking example is the "invisibility cloak" demonstrated by Pendry et al., in 2006. With the advancement of metamaterials, a wide range of optical devices have been realized through the method of transformation optics, including beam shifters, bent waveguides, beam splitters, Luneberg lenses, dielectric cloaks, and carpet cloaks [58–63]. However, the realization of artificial metameterials suffers from complex design and fabrication processes. It is difficult to construct transformation optical devices with large object size by using metamaterials. The operating wavelengths of these solid transformation optical devices are limited by the feature sizes of their nano-structures. It is difficult to operate at the visible light band. In optofluidic systems, the GRIN liquid media formed by diffusion between miscible flows at low-Pe level has the potential to be a new kind of material system. And it provides a versatile tool for designing transformation optical devices with controllable and spatially changing optical properties.

Based on the concepts above, researchers have used liquid flows as a new tunable transformation optics (TO) medium in optofluidic devices to manipulate light. Various fancy optofluidic TO devices were designed with novel optical properties. One of the most representative and fundamental designs was the liquid TO waveguide [64,65], and examples are shown in Figure 6. Compared with conventional liquid waveguides with high flow rates or high *Pe*, the optofluidic TO waveguide made full use of liquid diffusion between three flow streams to generate an inhomogeneous GRIN field in the transverse direction and the bidirectional direction, shown in Figure 6a. By changing the flow rates in a single liquid waveguide, spatially variable optical properties will be created to support novel optical phenomena such as self-focusing and interference [64], as shown in Figure 6b. The interference pattern in this optofluidic TO waveguide was similar to that in discrete diffraction. In addition, traditional discrete diffraction is usually generated in solid waveguide arrays whose feature sizes were several micrometers. It is fantastic that one can create a typical interference pattern through a simple liquid waveguide rather than solid waveguide arrays fabricated by complex processes. Furthermore, this liquid waveguide also possessed high tunability. As shown in Figure 6c, the light trajectory and converging points differed as the *Pe* decreased from 0.0100 to 0.005. The focusing period also increased because of a GRIN along with the direction of flow stream. In a relatively high *Pe* (0.07), the light traveled by straight lines. The relationship between the first section lengths and the flow rate of the core fluid in different boundary ratio was also shown in Figure 6c. In a later design, Yang et al., introduced a transformation Y-branch splitter by using an ethylene glycol solution as the high *RI* (*n* = 1.432) cladding flows and deionized water as the low *RI* (*n* = 1.333) core flow [65]. The RI-profile in this Y-branch splitter is bi-directional, as shown in Figure 6d. The flow rates of each input channel were calculated referring to the coordinate transformation for accurate light splitting. As a result, a wide range of split angles, from 0° to 30°, was achieved by choosing the proper flow rates, as shown in Figure 6e.

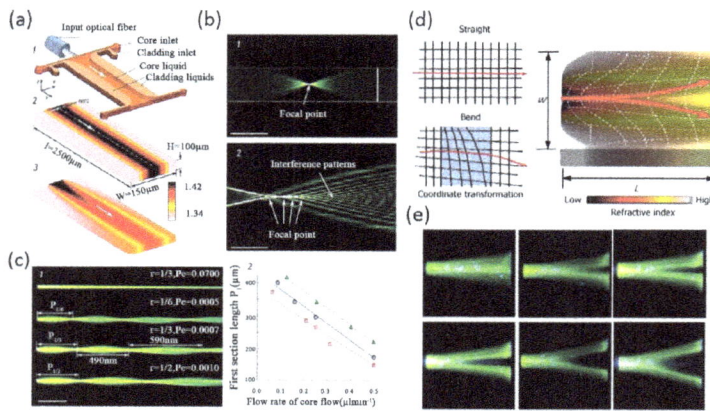

Figure 6. (**a**) Design of the optofluidic waveguides via a transformation optics approach; (**b**) Light focusing and interference in an optofluidic waveguide. The light trajectory and converging points as a function of the Pe and the flow rate (**c**); (**d**) Schematic illustration of the pure liquid optofluidic Y-branch splitter; (**e**) A wide range of split angle, from 0° to 30°, can be achieved by choosing the proper flow rates. Images reproduced with permission from [64,65].

It is unique that the convection–diffusion equation at low *Pe* shares the same mathematical form with quasi-conformal transformation optics (QCTO). Based on the concept of transformation optics not only have transformation optofluidic Y-branch splitters and waveguides been demonstrated, various devices like tunable waveguide bends and tunable liquid visible cloaking devices have also been designed for light manipulation by controlling the diffusion of liquid [66,67]. Figure 7a shows a tunable liquid TO waveguide bend, developed by Liu et al., The bend was formed by choosing the special boundary conditions of the diffusion process between ethylene glycol and deionized water. A gradient refractive index profile that coincided with that of the TO bend waveguide was achieved, thus steering the light path without optical loss in the same way as other TO systems [66]. The light beam profiles at the input and the output of the 90° bend and the 180° bend were almost the same, which means that the light beam maintained perfectly through the TO liquid waveguide. The manner of light in GRIN liquid waveguide bends was the same as that in homogenous straight liquid waveguides. Zhu et al., also creatively designed a tunable visible cloak by liquid diffusion [67], as shown in Figure 7b. A bump was designed at the bottom of the main channel to hide objects. It was easy to change the working states of this liquid visible cloak by controlling the motion of the three inlet miscible flows.

When the RI profile in the main channel mismatched with that of the QCTO, the incoming light is scattered by the bump. The liquid visible cloak was in a "cloak-off" state. As a result, the object behind the bump could be detected. In contrast, if the miscible flows were injected at enough low flow rates, an inhomogeneous *RI* profile showing analogy with that of QCTO generates. The liquid cloak maintained a "cloak- on" state. The reflected light ray in this device was the same as that in a flat mirror so that objects inside the bump were hidden.

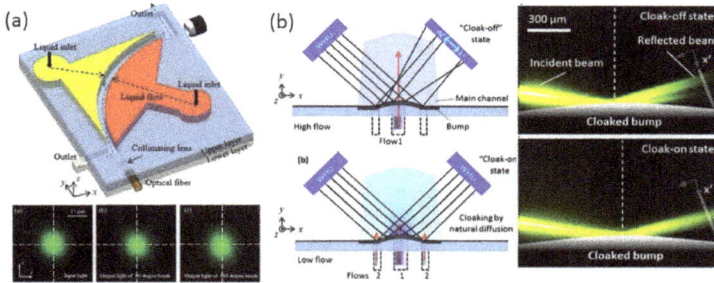

Figure 7. (a) Main concept of liquid waveguide bends and the light beam profiles of the liquid bends; (b) The "cloak-off" and "cloak-on" state of the switchable optofluidic carpet cloak using miscible liquids. Images reproduced with permission from [66,67].

4.2. Heat Conduction

The refractive index of a liquid is almost always related to its temperature. The temperature distribution through the heat conduction or thermal diffusion between several fluids in a microchannel with different temperatures will create a gradient refractive index profile similar to mass diffusion. This GRIN profile of the liquids in return will guide the propagation of the light beam [68]. Compared with traditional mass diffusion, the thermal diffusion can form an inhomogeneous GRIN profile in a homogeneous liquid flow. The use of a homogeneous liquid flow or single common liquid may simplify the liquid recycling process. In addition, thermal diffusion is more rapid than mess diffusion, which offers an opportunity for reducing the responding time of the optofluidic system. But these optofluidic devices based on heat conduction suffer from a relatively low *RI* difference, which limits the capability of manipulating light beams. Besides, an extra thermal field is needed to maintain the sustained inputs of liquids with a specific temperature different from room temperature.

The temperature distribution across the channel is controlled by the heat conduction equation:

$$\frac{\partial T}{\partial t} = \psi \nabla^2 T - U \nabla T \tag{9}$$

where T is the temperature, U is the average velocity of the liquid in microchannel. $\psi = k/\rho C_p$ is the thermal diffusivity, k is the thermal conductivity, ρ is the density of liquid, and C_p is the specific heat. The adjusting of RI can be realized by changing the temperature. According to the thermo-optics effect, the relationship between RI of a liquid and the temperature is expressed by

$$n(T) = n_0 + \varepsilon(T - T_0) \tag{10}$$

where n_0 is the RI at the initial temperature , and ε is the thermal coefficient of the liquid.

Tang et al., describes the design of a liquid thermal optical waveguide [65], as shown in Figure 8a. The refractive index of a liquid usually keeps negative correlation with the temperature. A flow stream with a lower temperature (21 °C) was sandwiched by two flow streams with a higher temperature (range from 30 °C to 80 °C). The heat conduction between the core flow and cladding flows results in a gradient RI distribution across the microchannel. This design enables the control of heat diffusion and then the RI by controlling the initial temperature difference and the flow rates in each input channel. In a later design, an optofluidic lens based on a laser-induced thermal gradient was demonstrated by Zhang et al., as shown in Figure 8b [69]. Compared with Tang's method, this approach was realized by an extra optical field instead of inserting liquids with different temperatures. Two straight chromium strips were fabricated at the bottom of the channel to absorb the energy of a pump laser. In this liquid thermal lens, benzyl alcohol solution was used because a relatively larger refractive index change can be obtained compared with other liquids such as water under a certain temperature difference. A 2D refractive index gradient will be formed between the two hot strips. It is demonstrated that the focal length can be continuously tuned from infinite to 1.3 mm. At the same time, an off-axis focusing can be realized by offsetting the heat spot of pump laser. This tunable lens possesses many advantages, such as small size, easy integration, and fast responding speed. However, it requires a more complex fabrication process and an extra optical field than previous thermal lenses. The efficiency of the laser-induced heating process is relatively low. Liu et al., also reported a liquid thermal GRIN lens in homogeneous fluids [70]. The focal length of the thermal lens can be adjusted from 500 μm to 430 μm. In this design, a relatively high enhancement factor can be achieved (5.4). And the corresponding full width at half maximum was 4 μm.

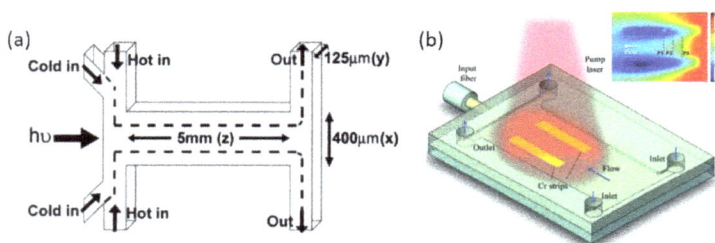

Figure 8. (**a**) Schematic design of the liquid thermal GRIN optical waveguides; (**b**) Schematic diagram of the optofluidic tunable lenses using laser-induced thermal gradient. Images reproduced with permission from [68,69].

5. Application in Biochemical Sensing

The capabilities of light manipulation and biochemical sensing are inherent along with the emergence of the optofluidics. The liquids in optofluidic systems are natural carriers for biological samples (i.e., cells, prokaryotes, DNA), nanoparticles, molecules like phosphate, and other water-soluble components. Optofluidics has the advantage of being highly-integrated, low-cost, fast in the field of biochemical sensing, detection and particle manipulation [12,13,15,16]. Optofluidics makes full use of the powerful tools and techniques in optics, such as evanescent wave fields, optical tweezers, and resonant cavity to enhance the function and efficiency of traditional microfluidic systems [71–75].

An evanescent wave field is one of the most widely used technologies in the detection of single nanoparticles/molecules with high sensitivity and signal-to-noise ratio. However, only a few samples in liquid can be illustrated by the evanescent wave at the solid-liquid interface. The samples are detected randomly. In order to overcome these drawbacks, one can either increase the intensity and the penetration depth of the evanescent field or confine the samples within the area illustrated by the evanescent wave. On the basis of these two approaches, Liang et al. demonstrated an optofluidic chip for nanoparticle detection [76], shown in Figure 9a. A silicone oil and paraffin oil mixture with high-RI was used as the sheath flow. While the low-RI ethylene glycol solution was used as the core flow. The interaction between the two immiscible fluid flows will generate an optically smooth and step-index interface, which was suitable for total internal reflection. By choosing the proper parameter of the RI and incident angle, the penetration depth can be broadened up to 1 μm as shown in Figure 9b. The sample core flow was then focused with a width narrower than 1 μm through the method of hydrodynamic focusing. This optofluidic chip realized the detection of every sample in the core flow through an evanescent wave without any failure. The application of TIR into optofludic systems will promote the improvement of detection systems with real-time control and rapid responses.

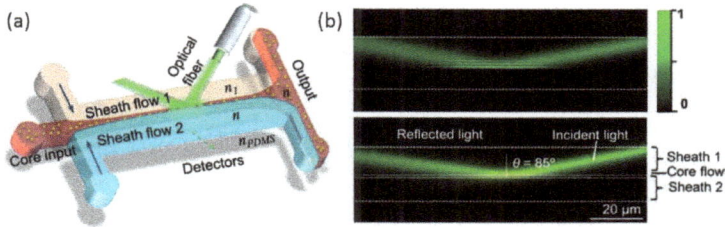

Figure 9. (**a**) Design of the optofluidic chip for single nanoparticle detection; (**b**) Optical intensity distribution of the evanescent field. Images reproduced with permission from [76].

Particle manipulation and sorting is another main application of optofluidcs. The optical tweezer is an efficient and effective tool for trapping micro particles by applying a strong focused laser beam [77]. Optical tweezers, especially holographic optical tweezers (HOTs) have been applied in the fields of biology, optical manipulation, and channel-facilitated diffusion [78–80]. By combining the microfluidic array and HOTs creatively, researchers have succeeded in investigating single-file diffusion of Brownian particles [81]. A solid objective lens is usually applied to form optical tweezers in conventional approaches. As shown in Figure 10a, based on the unique optical properties of the thermal GRIN lens, Liu et al., realized the trapping of a single living cell in a dynamic liquid environment [70]. By varying the focus length of the optofluidic device, the living cell can be trapped at different locations, which provides an approach for the manipulation and analysis of a single living cell. Wu et al., combined the optical force and the opposite impinging streams to achieve the size-selective optical sorting of gold nanoparticles in fluids [82], as shown in Figure 10b. The injection of the two opposite laminar streams meeting at the junction generate a smooth stagnation point, which will decelerate the moving nanoparticles. In other words, this design of the impinging streams can prolong the function time of optical force. In return, this optofluidic sorter for NPs owns higher efficiency than other sorting methods, such as centrifugation, electrophoresis, and size exclusion etc. In the experiment, the sorting of different-sized nanoparticles is demonstrated successfully. The sorting efficiency for 50/100 nm and 100/200 nm mixtures are 92% and 86%, respectively. A sorting output of 300 particles per minute was realized.

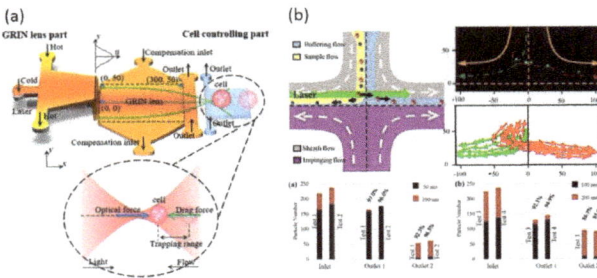

Figure 10. (**a**) The design of the liquid thermal GRIN lens for trapping living cells in a dynamic liquid environment; (**b**) Precise sorting of gold nanoparticles in a flowing system where the sorting efficiency is as high as 92%. Images reproduced with permission from [70,82].

In addition, several optofluidic chips were demonstrated to measure the effective refractive index of living cells through novel optical techniques such as Fabry–Pérot (FP) resonant cavity, fiber Bragg grating resonant cavity, and the Mach–Zehnder interferometer etc. [83–85]. The resolution of the refractive index unit (RIU) can reach the order of 10^{-3}. A typical cell refractive index model is shown in Figure 11a,b. Aside from applications in particle and cell sensing and manipulation, optofluidics also finds important applications in environmental detection. By the combination of fluid control and optical detection in the scale of micron meters, optofluidic biochemical devices could be designed with small size, low cost, parallel processing, and real-time monitoring. Zhu et al., reported a lab-on-a-chip analysis system for phosphate detection [86]. A FP resonant cavity was fabricated by two opposite aligned Au-coated fibers to enhance the absorption of phosphate. Compared with traditional spectroscopy instruments, the optofluidic phosphate detector design possesses several superiorities such as miniaturization with short absorption cell length down to 300 μm and fast detection (6 s).

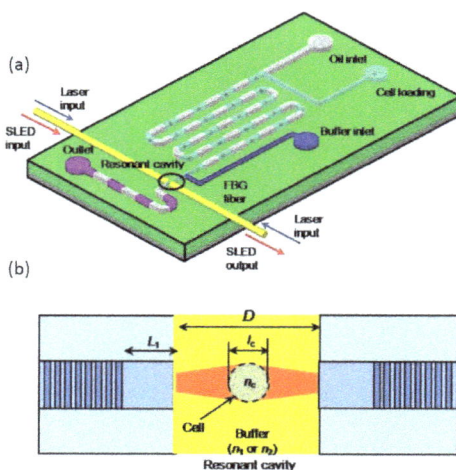

Figure 11. (**a**) Schematic diagram of the biochip design and (**b**) the fiber Bragg grating resonant cavity. Images reproduced with permission from [85].

6. Summary and Outlook

This paper reviews optofluidic lab-on-a-chip techniques based on an inhomogeneous liquid flow for light manipulation and demonstrates some application examples in biochemical sensing. Optofluidic lab-on-chip manipulation techniques and designs are categorized according to the interaction between different flow streams. Emblematical and significant works are introduced from the perspectives of manipulation of a liquid-liquid interface and that of the liquid gradient refractive index. In these woks, researchers find ways to control the liquid in a microchannel, realizing light routing, bending, switching, focusing, and interference etc. Fluids can be easily reconfigured and replaced, allowing for much larger tunability in the refractive index and flexibility in shape than solid equivalents. By manipulating flow rate and liquid compositions, the function of the optofluidic light manipulation devices or systems can be fully exploited. Besides the tunability and reconfigurability, optofluidic devices possess another outstanding advantage, easy integration. Compared with conventional optical systems, optofluidic devices can be fabricated and integrated in other MEMS chips as an optical control element.

In the future, researchers may focus on new microfluidic liquid control methods and potential techniques to improve the performance of optofluidic systems. Recently, transformation optics has drawn a lot of intention. Researchers will come across new propositions and research points after applying the concepts and designing new methods of transformation optics in optofluidics. As a result, the marriage of transformation optics and optofludics will promote various novel qualities in light manipulation. No matter what the future will be, the optofluidic lab-on-a-chip system based on pure liquid is a powerful concept for light manipulation. Based on this, increasing optofluidic systems or devices for real-time monitoring with properties of fast, accurate, low-cost, small-sized biochemical micro- sensors will be brought into existence.

Acknowledgments: This work was financially supported by the National Natural Science Foundation of China (No. 11774274, 61378093), Open Foundation of National Laboratory for Marine Science and Technology (No. QNLM2016ORP0410) and State Oceanic Administration, the People's Republic of China "marine environmental monitoring and upgrading". We also acknowledge the Center for Nanoscience and Nanotechnology at Wuhan University for providing assistance with nanofabrication. Finally, the author wants to thank Yu Gao for careful reading and revising.

Author Contributions: Yunfeng Zuo designed and wrote the paper; Yi Yang supervised the work; Xiaoqiang Zhu, Yang Shi and Li Liang revised the framework and English writing.

Conflicts of Interest: The authors declare no conflict of interest.

References

1. Erickson, D.; Heng, X.; Li, Z.Y.; Rockwood, T.; Emery, T.; Zhang, Z.Y.; Scherer, A.; Yang, C.H.; Psaltis, D. Optofluidics. *Proc. SPIE* **2005**, *5908*, 59080S-1.

2. Whitesides, G.M. The origins and the future of microfluidics. *Nature* **2006**, *442*, 368–373. [CrossRef] [PubMed]

3. Zhao, W.D.; Wang, B.J.; Wang, W. Biochemical sensing by nanofluidic crystal in a confined space. *Lab Chip* **2016**, *16*, 2050–2058. [CrossRef] [PubMed]

4. Wang, N.; Zhang, X.M.; Wang, Y.; Yu, W.X.; Chan, H.L. Microfluidic reactors for photocatalytic water purification. *Lab Chip* **2014**, *14*, 1074–1082. [CrossRef] [PubMed]

5. Huang, X.W.; Liu, J.; Yang, Q.J.; Liu, Y.; Zhu, Y.J.; Li, T.H.; Tsang, Y.H.; Zhang, X.M. Microfluidic chip-based one-step fabrication of artificial photosystem I for photocatalytic cofactor regeneration. *RSC Adv.* **2016**, *6*, 101974–101980. [CrossRef]

6. Shui, L.L.; Pennathur, S.; Eijkel, J.C.T. Multiphase flow in lab on chip devices: A real tool for the future. *Lab Chip* **2008**, *8*, 1010–1014. [PubMed]

7. Xie, Y.B.; Bos, D.; Vreede, L.J.D.; Boer, H.L.D.; Meulen, M.J.V.D.; Versluis, M.; Sprenkels, A.J.; Berg, A.V.D.; Eijkel, J.C.T. High-efficiency ballistic electrostatic generator using microdroplets. *Nat. Commun.* **2014**, *5*, 3575. [CrossRef] [PubMed]

8. Gossett, D.R.; Weaver, W.M.; Mach, A.J.; Hur, S.C.; Tse, H.T.K.; Lee, W.; Amini, H.; Carlo, D.D. Label-free cell separation and sorting in microfluidic systems. *Anal. Bioanal. Chem.* **2010**, *397*, 3249–3267. [CrossRef] [PubMed]

9. Gossett, D.R.; Tse, H.T.K.; Lee, S.A.; Ying, Y.; Lindgen, A.G.; Yang, O.O.; Rao, J.Y.; Clark, A.T.; Carlo, D.D. Hydrodynamic stretching of single cells for large population mechanical phenotyping. *Proc. Natl. Acad. Sci. USA* **2012**, *109*, 7630–7635. [CrossRef] [PubMed]

10. Pagliara, S.; Franze, K.; McClain, C.R.; Wylde, G.W.; Fisher, C.L.; Franklin, R.J.M.; Kabla, A.J.; Keyser, U.F.; Chalut, K.J. Auxetic nuclei in embryonic stem cells exiting pluripotency. *Nat. Mater.* **2014**, *13*, 638–644. [CrossRef] [PubMed]

11. Zilionis, R.; Nainys, J.; Veres, A.; Savova, V.; Zemmour, D.; Klein, A.M.; Mazutis, L. Single-cell barcoding and sequencing using droplet microfluidics. *Nat. Protoc.* **2017**, *12*, 44–73. [CrossRef] [PubMed]

12. Psaltis, D.; Quake, S.R.; Yang, C.H. Developing optofluidic technology through the fusion of microfluidics and optics. *Nature* **2006**, *442*, 381–386. [CrossRef] [PubMed]

13. Minzioni, P.; Osellame, R.; Sada, C.; Zhao, S.; Omenetto, F.G.; Gylfason, K.B.; Haraldsson, T.; Zhang, Y.; Ozcan, A.; Wax, A.; et al. Roadmap for optofluidics. *J. Opt.* **2017**, *19*, 093003. [CrossRef]

14. Schmidt, H.; Hawkins, A.R. The photonic integration of non-solid media using optofluidics. *Nat. Photonics* **2011**, *5*, 598–604. [CrossRef]

15. Erickson, D.; Sinton, D.; Psaltis, D. Optofluidics for energy applications. *Nat. Photonics* **2011**, *5*, 583–590. [CrossRef]

16. Fan, X.D.; White, I.M. Optofluidic microsystems for chemical and biological analysis. *Nat. Photonics* **2011**, *5*, 591–597. [CrossRef] [PubMed]

17. Zhao, Y.H.; Stratton, Z.S.; Guo, F.; Lapsley, M.L.; Chan, C.Y.; Lin, S.S.C.; Huang, T.J. Optofluidic imaging: Now and beyond. *Lab Chip* **2013**, *13*, 17–24. [CrossRef] [PubMed]

18. Pagliara, S.; Camposeo, A.; Polini, A.; Cingolani, R.; Pisignano, D. Electrospun light-emitting nanofibers as excitation source in microfluidic devices. *Lab Chip* **2009**, *9*, 2851–2856. [CrossRef] [PubMed]

19. Salafi, T.; Zeming, K.K.; Zhang, Y. Advancements in microfluidics for nanoparticle separation. *Lab Chip* **2017**, *17*, 11–33. [CrossRef] [PubMed]

20. Pagliara, S.; Chimerel, C.; Langford, R.; Aarts, D.G.A.L.; Keyser, U.F. Parallel sub-micrometre channels with different dimensions for laser scattering detection. *Lab Chip* **2011**, *11*, 3365–3368. [CrossRef] [PubMed]

21. Lee, W.; Kwon, D.; Choi, W.; Jung, G.Y.; Au, A.K.; Folch, A.; Jeon, S. 3D-Printed microfluidic device for the detection of pathogenic bacteria using size-based separation in helical channel with trapezoid cross-section. *Sci. Rep.* **2015**, *5*, 7717. [CrossRef] [PubMed]

22. Cama, J.; Chimerel, C.; Pagliara, S.; Javer, A.; Keyser, U.F. A label-free microfluidic assay to quantitatively study antibiotic diffusion through lipid membranes. *Lab Chip* **2014**, *14*, 2303–2308. [CrossRef] [PubMed]

23. Schmidt, H.; Aaron, R.; Hawkins, A.R. Optofluidic waveguides: I. Concepts and implementations. *Microfluid. Nanofluid.* **2008**, *4*, 3–16. [CrossRef] [PubMed]

24. Fei, P.; He, Z.; Zheng, C.; Chen, T.; Men, Y.; Huang, Y. Discretely tunable optofluidic compound microlenses. *Lab Chip* **2011**, *11*, 2835–2841. [CrossRef] [PubMed]

25. Nguyen, N.T. Micro-optofluidic Lenses: A review. *Biomicrofluidics* **2010**, *4*, 031501. [CrossRef] [PubMed]

26. Xiong, S.; Liu, A.Q.; Chin, L.K.; Yang, Y. An optofluidic prism tuned by two laminar flows. *Lab Chip* **2011**, *11*, 1864–1869. [CrossRef] [PubMed]

27. Chao, K.S.; Lin, M.S.; Yang, R.J. An in-plane optofluidic microchip for focal point control. *Lab Chip* **2013**, *13*, 3886–3892. [CrossRef] [PubMed]

28. Yu, J.Q.; Yang, Y.; Liu, A.Q.; Chin, L.K.; Zhang, X.M. Microfluidic droplet grating for reconfigurable optical diffraction. *Opt. Lett.* **2010**, *35*, 1890–1892. [CrossRef] [PubMed]

29. Chin, L.K.; Liu, A.Q.; Soh, Y.C.; Lim, C.S.; Lin, C.L. A reconfigurable optofluidic Michelson interferometer using tunable droplet grating. *Lab Chip* **2010**, *10*, 1072–1078. [CrossRef] [PubMed]

30. Seow, Y.C.; Lim, S.P.; Lee, H.P. Tunable optofluidic switch via hydrodynamic control of laminar flow rate. *Appl. Phys. Lett.* **2009**, *95*, 114105. [CrossRef]

31. Song, W.Z.; Psaltis, D. Pneumatically tunable optofluidic 2×2 switch for reconfigurable optical circuit. *Lab Chip* **2011**, *11*, 2397–2402. [CrossRef] [PubMed]

32. Song, W.Z.; Vasdekis, A.E.; Li, Z.Y.; Psaltis, D. Low-order distributed feedback optofluidic dye laser with reduced threshold. *Appl. Phys. Lett.* **2009**, *94*, 051117. [CrossRef]

33. Chen, Y.; Lei, L.; Zhang, K.; Shi, L.; Wang, L.; Li, H.; Zhang, X.M.; Wang, Y.; Chan, H.L. Optofluidic microcavities: Dye-lasers and biosensors. *Biomicrofluidics* **2010**, *4*, 043002. [CrossRef] [PubMed]

34. Squires, T.M. Microfluidics: Fluid physics at the nanoliter scale. *Rev. Mod. Phys.* **2005**, *77*, 977–1026. [CrossRef]

35. Mala, G.M.; Li, D. Flow characteristics of water in microtubes. *Int. J. Heat Fluid Flow* **1999**, *20*, 142–148. [CrossRef]

36. Carlo, D.D. Inertial microfluidics. *Lab Chip* **2009**, *9*, 3038–3046. [CrossRef] [PubMed]

37. Perumal, M.; Ranga Raju, K.G. Approximate convection-diffusion equations. *J. Hydrol. Eng.* **1999**, *4*, 160–164. [CrossRef]

38. Pan, M.; Kim, M.; Kuiper, S.; Tang, S.K.T. Actuating fluid–fluid interfaces for the reconfiguration of light. *IEEE J. Sel. Top. Quantum Electron.* **2015**, *21*, 9100612. [CrossRef]

39. Lee, G.B.; Chang, C.C.; Huang, S.B.; Yang, R.J. The hydrodynamic focusing effect inside rectangular microchannels. *J. Micromech. Microeng.* **2006**, *16*, 1024–1032. [CrossRef]

40. Wu, Z.G.; Nguyen, N.T. Hydrodynamic focusing in microchannels under consideration of diffusive dispersion: Theories and experiments. *Sens. Actuators B* **2005**, *107*, 965–974. [CrossRef]

41. Knight, J.B.; Vishwanath, A.; Brody, J.P.; Austin, R.H. Hydrodynamic focusing on a silicon chip: Mixing nanoliters in microseconds. *Phys. Rev. Lett.* **1998**, *80*, 3863–3866. [CrossRef]

42. Wolfe, D.B.; Conroy, R.S.; Garstecki, P.; Mayers, B.T.; Fischbach, M.A.; Paul, K.E.; Prentiss, M.; Whitesides, G.M. Dynamic control of liquid-core/liquid-cladding optical waveguides. *Proc. Natl. Acad. Sci. USA* **2004**, *101*, 12434–12438. [CrossRef] [PubMed]

43. Tang, S.K.Y.; Stan, C.A.; Whitesides, G.M. Dynamically reconfigurable liquid-core liquid-cladding lens in a microfluidic channel. *Lab Chip* **2008**, *8*, 395–401. [CrossRef] [PubMed]

44. Seow, Y.C.; Liu, A.Q.; Chin, L.K.; Li, X.C.; Huang, H.J.; Cheng, T.H.; Zhou, X.Q. Different curvatures of tunable liquid microlens via the control of laminar flow rate. *Appl. Phys. Lett.* **2008**, *93*, 084101. [CrossRef]

45. Song, C.L.; Nguyen, N.T.; Asundi, A.K.; Low, C.L.N. Biconcave micro-optofluidic lens with low-refractive-index liquids. *Opt. Lett.* **2009**, *34*, 3622–3624. [CrossRef] [PubMed]

46. Mao, X.L.; Waldeisen, J.R.; Huang, T.J. "Microfluidic drifting"—Implementing three-dimensional hydrodynamic focusing with a single-layer planar microfluidic device. *Lab Chip* **2007**, *7*, 1260–1262. [CrossRef] [PubMed]

47. Lee, K.S.; Kim, S.B.; Lee, K.H.; Sung, H.J.; Kim, S.S. Three-dimensional microfluidic liquid-core/liquid-cladding waveguide. *Appl. Phys. Lett.* **2010**, *97*, 021109. [CrossRef]

48. Mao, X.L.; Waldeisen, J.R.; Juluri, B.K.; Huang, T.J. Hydrodynamically tunable optofluidic cylindrical microlens. *Lab Chip* **2007**, *7*, 1303–1308. [CrossRef] [PubMed]

49. Yang, Y.; Liu, A.Q.; Lei, L.; Chin, L.K.; Ohl, C.D.; Wang, Q.J.; Yoon, H.S. A tunable 3D optofluidic waveguide dye laser via two centrifugal Dean flow streams. *Lab Chip* **2011**, *11*, 3182–3187. [CrossRef] [PubMed]

50. Rosenauer, M.; Vellekoop, M.J. 3D fluidic lens shaping—A multiconvex hydrodynamically adjustable optofluidic microlens. *Lab Chip* **2009**, *9*, 1040–1042. [CrossRef] [PubMed]

51. Li, L.; Zhu, X.Q.; Liang, L.; Zuo, Y.F.; Xu, Y.S.; Yang, Y.; Yuan, Y.J.; Huang, Q.Q. Switchable 3D optofluidic Y-branch waveguides tuned by Dean flows. *Sci. Rep.* **2016**, *6*, 38338. [CrossRef] [PubMed]

52. Liang, L.; Zhu, X.Q.; Liu, H.L.; Shi, Y.; Yang, Y. A switchable 3D liquid–liquid biconvex lens withenhanced resolution using Dean flow. *Lab Chip* **2017**, *17*, 3258–3263. [CrossRef] [PubMed]

53. Wolfe, D.B.; Vezenov, D.V.; Mayers, B.T.; Whitesides, G.M.; Conroy, R.S.; Prentiss, M.G. Diffusion-controlled optical elements for optofluidics. *Appl. Phys. Lett.* **2005**, *87*, 181105. [CrossRef]

54. Mao, X.L.; Steven Lin, S.Z.; Lapsley, M.I.; Shi, J.J.; Juluri, B.K.; Huang, T.J. Tunable liquid gradient refractive index (L-GRIN) lens with two degrees of freedom. *Lab Chip* **2009**, *9*, 2050–2058. [CrossRef] [PubMed]

55. Zhao, H.T.; Yang, Y.; Chin, L.K.; Chen, H.F.; Zhu, W.M.; Zhang, J.B.; Yap, P.H.; Liedberg, B.; Wang, K.; Wang, G.; et al. Optofluidic lens with low spherical and low field curvature aberrations. *Lab Chip* **2016**, *16*, 1617–1624. [CrossRef] [PubMed]

56. Shi, Y.; Liang, L.; Zhu, X.Q.; Zhang, X.M.; Yang, Y. Tunable self-imaging effect using hybrid optofluidic waveguides. *Lab Chip* **2015**, *15*, 4398–4403. [CrossRef] [PubMed]

57. Chen, H.Y.; Chan, C.T.; Sheng, P. Transformation optics and metamaterials. *Nat. Mater.* **2010**, *9*, 387–396. [CrossRef] [PubMed]

58. Roberts, D.A.; Rahm, M.; Pendry, J.B.; Smith, D.R. Transformation-optical design of sharp waveguide bends and corners. *Appl. Phys. Lett.* **2008**, *93*, 251111. [CrossRef]

59. Huangfu, J.T.; Xi, S.; Kong, F.M.; Zhang, J.J.; Chen, H.S.; Wang, D.X.; Wu, B.L.; Ran, L.X.; Kong, J.A. Application of coordinate transformation in bent waveguides. *J. Appl. Phys.* **2008**, *104*, 014502. [CrossRef]

60. Li, J.S.; Pendry, J.B. Hiding under the carpet: A new strategy for cloaking. *Phys. Rev. Lett.* **2008**, *101*, 203901. [CrossRef] [PubMed]

61. Ergin, T.; Stenger, N.; Brenner, P.; Pendry, J.B.; Wegener, M. Three-dimensional invisibility cloak at optical wavelengths. *Science* **2010**, *328*, 337–339. [CrossRef] [PubMed]

62. Liu, R.; Ji, C.; Mock, J.J.; Chin, J.Y.; Cui, T.J.; Smith, D.R. Broadband ground-plane cloak. *Science* **2009**, *323*, 366–369. [CrossRef] [PubMed]

63. Rahm, M.; Cummer, S.A.; Schurig, D.; Pendry, J.B.; Smith, D.R. Optical design of reflectionless complex media by finite embedded coordinate transformations. *Phys. Rev. Lett.* **2008**, *100*, 063903. [CrossRef] [PubMed]

64. Yang, Y.; Liu, A.Q.; Chin, L.K.; Zhang, X.M.; Tsai, D.P.; Lin, C.L.; Lu, C.; Wang, G.P.; Zheludev, N.I. Optofluidic waveguide as a transformation optics device for lightwave bending and manipulation. *Nature Commun.* **2012**, *3*, 651. [CrossRef] [PubMed]

65. Yang, Y.; Chin, L.K.; Tsai, J.M.; Tsai, D.P.; Zheludev, N.I.; Liu, A.Q. Transformation optofluidics for large-angle light bending and tuning. *Lab Chip* **2012**, *12*, 3785–3790. [CrossRef] [PubMed]

66. Liu, H.L.; Zhu, X.Q.; Liang, L.; Zhang, X.M.; Yang, Y. Tunable transformation optical waveguide bends in liquid. *Optica* **2017**, *4*, 839–846. [CrossRef]
67. Zhu, X.Q.; Liang, L.; Zuo, Y.F.; Zhang, X.M.; Yang, Y. Tunable visible cloaking using liquid diffusion. *Laser Photonics Rev.* **2017**, *11*, 1700066. [CrossRef]
68. Tang, S.K.T.; Mayers, B.T.; Vezenov, D.V.; Whitesides, G.M. Optical waveguiding using thermal gradients across homogeneous liquids in microfluidic channels. *Appl. Phys. Lett.* **2006**, *88*, 061112. [CrossRef]
69. Chen, Q.M.; Jian, A.Q.; Li, Z.H.; Zhang, X.M. Optofluidic tunable lenses using laser-induced thermal gradient. *Lab Chip* **2016**, *16*, 104–111. [CrossRef] [PubMed]
70. Liu, H.L.; Shi, Y.; Liang, L.; Li, L.; Guo, S.S.; Yin, L.; Yang, Y. A liquid thermal gradient refractive index lens and sing it to trap single living cell in flowing environments. *Lab Chip* **2017**, *17*, 1280–1286. [CrossRef] [PubMed]
71. Özbakır, Y.; Jonáš, A.; Kiraz, A.; Erkey, C. Total internal reflection-based optofluidic waveguides fabricated in aerogels. *J. Sol-Gel Sci. Technol.* **2017**, *84*, 522–534. [CrossRef]
72. Liu, P.Y.; Chin, L.K.; Ser, W.; Chen, H.F.; Hsieh, C.M.; Lee, C.H.; Sung, K.B.; Ayi, T.C.; Yap, P.H.; Liedberg, B.; et al. Cell refractive index for cell biology and disease diagnosis: Past, present and future. *Lab Chip* **2016**, *16*, 634–644. [CrossRef] [PubMed]
73. Cho, S.H.; Godin, J.M.; Chen, C.H.; Qiao, W.; Lee, H.; Lo, Y.H. Review Article: Recent advancements in optofluidic flow cytometer. *Biomicrofluidics* **2010**, *4*, 043001. [CrossRef] [PubMed]
74. Domachuk, P.; Cronin-Golomb, M.; Eggleton, B.J. Application of optical trapping to beam manipulation in optofluidics. *Opt. Express* **2005**, *13*, 7265–7275. [CrossRef] [PubMed]
75. Yang, A.H.; Moore, S.D.; Schmidt, B.S.; Klug, M.; Lipson, M.; Erickson, D. Optical manipulation of nanoparticles and biomolecules in sub-wavelength slot waveguides. *Nature* **2009**, *457*, 71–75. [CrossRef] [PubMed]
76. Liang, L.; Zuo, Y.F.; Wu, W.; Zhu, X.Q.; Yang, Y. Optofluidic restricted imaging, spectroscopy and counting of nanoparticles by evanescent wave using immiscible liquids. *Lab Chip* **2016**, *16*, 3007–3014. [CrossRef] [PubMed]
77. Ashkin, A. Optical trapping and manipulation of neutral particles using lasers. *Proc. Natl. Acad. Sci. USA* **1997**, *94*, 4853–4860. [CrossRef] [PubMed]
78. Padgett, M.; Leonardo, R.D. Holographic optical tweezers and their relevance to lab on chip devices. *Lab Chip* **2011**, *11*, 1196–1205. [CrossRef] [PubMed]
79. Grier, D.G. A revolution in optical manipulation. *Nature* **2003**, *424*, 810–816. [CrossRef] [PubMed]
80. Pagliara, S.; Dettmer, S.L.; Keyser, U.F. Channel-facilitated diffusion boosted by particle binding at the channel entrance. *Phys. Rev. Lett.* **2014**, *113*, 048102. [CrossRef] [PubMed]
81. Padgett, M.; Bowman, R. Tweezers with a twist. *Nat. Photonics* **2011**, *5*, 343–348. [CrossRef]
82. Wu, W.; Zhu, X.Q.; Zuo, Y.F.; Liang, L.; Zhang, S.P.; Zhang, X.M.; Yang, Y. Precise sorting of gold nanoparticles in a flowing system. *ACS Photonics* **2016**, *3*, 2497–2504. [CrossRef]
83. Song, W.Z.; Liu, A.Q.; Swaminathan, S.; Lim, C.S.; Yap, P.H.; Ayi, T.C. Determination of single living cell's dry/water mass using optofluidic chip. *Appl. Phys. Lett.* **2007**, *91*, 223902. [CrossRef]
84. Song, W.Z.; Zhang, X.M.; Liu, A.Q.; Lim, C.S.; Yap, P.H.; Hosseini, H.M.M. Refractive index measurement of single living cells using on-chip Fabry-Pérot cavity. *Appl. Phys. Lett.* **2006**, *89*, 203901. [CrossRef]
85. Chin, L.K.; Liu, A.Q.; Lim, C.S.; Zhang, X.M.; Ng, J.H.; Hao, J.Z.; Takahashi, S. Differential single living cell refractometry using grating resonant cavity with optical trap. *Appl. Phys. Lett.* **2007**, *91*, 243901. [CrossRef]
86. Zhu, J.M.; Shi, Y.; Zhu, X.Q.; Yang, Y.; Jiang, F.H.; Sun, C.J.; Zhao, W.H.; Han, X.T. Optofluidic marine phosphate detection with enhanced absorption using a Fabry–Pérot resonator. *Lab Chip* **2017**, *17*, 4025–4030. [CrossRef] [PubMed]

micromachines

MDPI

Review

Optofluidic Tunable Lenses for In-Plane Light Manipulation

Qingming Chen [1], Tenghao Li [1], Zhaohui Li [2], Jinlin Long [3] and Xuming Zhang [1,4,*]

[1] Department of Applied Physics, The Hong Kong Polytechnic University, Hong Kong 999077, China;
qing-ming.chen@connect.polyu.hk (Q.C.); 14900295r@connect.polyu.hk (T.L.)

[2] School of Electronics and Information Technology, Sun Yat-Sen University, Guangzhou 510275, China;
li_zhaohui@hotmail.com

[3] School of Chemistry and Chemical Engineering, Fuzhou University, Fuzhou 350116, China;
jllong@fzu.edu.cn

[4] Shenzhen Research Institute of the Hong Kong Polytechnic University, Shenzhen 518057, China

* Correspondence: apzhang@polyu.edu.hk; Tel.: +852-3400-3258

Received: 22 January 2018; Accepted: 11 February 2018; Published: 26 February 2018

Abstract: Optofluidics incorporates optics and microfluidics together to construct novel devices for microsystems, providing flexible reconfigurability and high compatibility. Among many novel devices, a prominent one is the in-plane optofluidic lens. It manipulates the light in the plane of the substrate, upon which the liquid sample is held. Benefiting from the compatibility, the in-plane optofluidic lenses can be incorporated into a single chip without complicated manual alignment and promises high integration density. In term of the tunability, the in-plane liquid lenses can be either tuned by adjusting the fluidic interface using numerous microfluidic techniques, or by modulating the refractive index of the liquid using temperature, electric field and concentration. In this paper, the in-plane liquid lenses will be reviewed in the aspects of operation mechanisms and recent development. In addition, their applications in lab-on-a-chip systems are also discussed.

Keywords: optofluidics; microfluidics; in-plane liquid lens; lab-on-a-chip

1. Introduction

Nowadays, miniaturized systems are playing important roles in both academic research and industrial applications. More elements and functions have been incorporated into a single chip, reducing the cost and improving the performance at the same time. Optofluidics combines optics and microfluidics together to construct novel elements for lab-on-a-chip applications [1–7]. By replacing the solid materials with liquids, it enables the flexible modulation of optical properties and improves compatibilities. Compared with its solid counterpart, optofluidics has some unique merits [8–11], such as flexible tunability, good compatibility, small size and easy fabrication, etc. It has been intensively studied by numerous research communities [1–3,7,12,13]. A number of applications have been achieved using the optofluidic techniques, such as chemical and biological detection [4,5,14], particle manipulation [15,16], optofluidic laser [13,17], tunable waveguides [18,19], and reconfigurable optofluidic lens [12]. Among them, the optofluidic lens is the most extensive exploited part. In conventional optical systems, the optical lens is used to change the propagation of light and reshape the beam. The solid lens has constant refractive index (RI) and a fixed focal length. The optical modulation is achieved by mechanical movement, which is complicated and poorly scalable. While the optofluidic lens enables easy modulation of the optical properties by either changing the lens geometry or tuning RI of the liquid.

By the manipulation of liquids in microscale, reconfigurable optofluidic lens has been demonstrated in a small chip. According to the propagation of the beam, the liquid lenses can be generally classified into two categories: out-of-plane lens and in-plane lens [12]. The former manipulates the beam in the direction perpendicular to the substrate of microfluidic chip [20], which is similar to the conventional lens. It can be used to replace the solid lens in some miniature optical systems that require variable focusing. The latter deals with the light in the direction parallel to the substrate of microfluidic chip, providing an effective way to manipulate the beam in a microfluidic network. Therefore, the sample and the probe beam can be delivered in the same liquid layer, enhancing the interaction of light and matter. Thus, it promises better scalability and more complexity for optofluidic networks. Some reviews have reported the development and applications of the optofluidic lenses [12,21–23]. Nam-Trung Nguyen gave a comprehensive review of the optofluidic lenses on their categories and working principles [12]. It focused on the schematic designs of different optofluidic lenses and some characteristic parameters, such as the response time and the tunability of focal length and RI. It also listed some liquids that are widely used in optofluidics. Mishra et al. reported the general characteristics and characterization methods of the optofluidic lenses, including the actuation methods and the spherical aberration [22]. In particular, Mishra had deeply discussed the aberration control, which is very important in optical imaging. Xu et al. discussed the development of dielectrophoretically tunable optofluidic lenses [23]. And Krogmann presented the design, fabrication and optical properties of the electrowetting based micro-optical components [24]. However, a comprehensive description of in-plane liquid lenses and their perspectives in optofluidic networks is still required. This review focuses on the in-plane optofluidic lenses, including the working mechanisms and their applications in lab-on-a-chip systems.

Figure 1a summarizes the categories of the in-plane lenses and Figure 1b explains the corresponding working principles. Most of the reported designs of in-plane optofluidic lenses can classified into two types: refractive lens and gradient index (GRIN) lens. The former often makes use of interfacial deformation and the latter of RI modulation. The rest of reported designs can be generally grouped into the others in this review. In the refractive lens, the beam is refracted at the smooth fluidic interface of immiscible liquids (see Figure 1(b1)). The focal length is usually tuned by changing the lens geometry. In a microfluidic chip, there are numerous ways to modify the curvature of the fluidic interface, for example, pressure control [25], hydrodynamic modulation [26,27], electrowetting [24] and dielectrophoresis [23]. Among them, the pressure control and the hydrodynamic modulation are more popular in regulating in-plane lenses. In the GRIN lens, solution diffusion [28] or thermal diffusion [29,30] can be used to establish a RI gradient profile, in which the rays are bent gradually and then focused to a point (Figure 1(b2)). A large RI gradient (usually over 0.1) is achievable in a microscale region, resulting in tight focusing and wide tunable range. The refractive lens and the GRIN lens are often independent of the polarization and the wavelength of the incident light due to the use of isotropic and low-dispersion liquids (e.g., water, ethanol, ethylene glycol). In addition, there are some other lenses that aim at polarization separation or wavelength selection, such as birefringent liquid lens [31] and diffractive optofluidic lens [32]. For example, liquid crystal (LC) has been used to construct a polarization-dependent liquid lens [31]. Figure 1(b3) displays the schematic of the LC liquid lens, in which the LC molecules can be realigned by applying a sufficiently strong electrical field. As a result, incident beams of different polarization directions would experience different RIs and have different focal lengths. Another kind of lens is Fresnel zone plate (FZP), which is based on the diffraction rather than refraction. As shown in Figure 1(b4), an optofluidic FZP can be achieved for light manipulation by filling liquid into the periodic microstructure. Different parts of the diffracted lights interference constructively with each other to form a focal point. The FZP depends on the incident wavelength, providing another freedom of tunability.

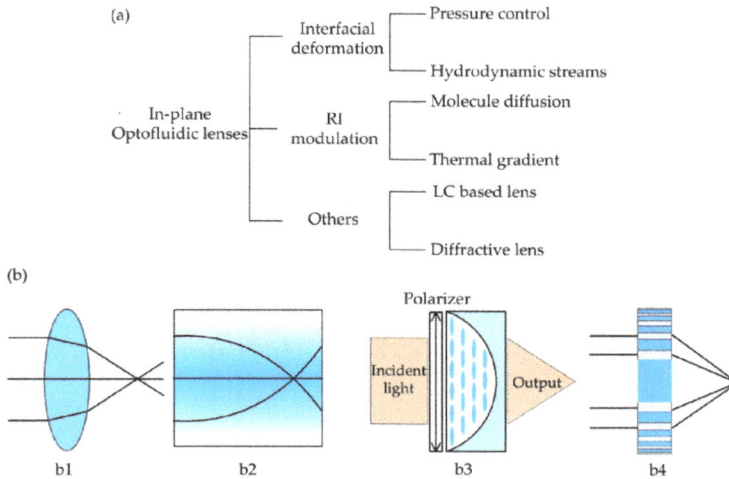

Figure 1. In-plane optofluidic lenses categories and the working principles: (**a**) The in-plane liquid lenses are classified into three types of lenses according to their working principles; (**b**) schematic diagrams of the in-plane liquid lens: (**b1**) is the interfacial deformation lens; (**b2**) is RI modulation (gradient index) lens; (**b3**) is the liquid-crystalbased lens (LC: the blue ellipses) and (**b4**) is the diffractive lens (i.e., the Fresnel zone plate).

This article has five sections. The first gives an introduction of optofluidic lenses. In the second, the in-plane optofluidic lenses will be discussed according to their operation mechanisms. Then, some applications of the in-plane liquid lenses will be presented in the third section. There is a brief discussion in section four. The last part is the conclusion.

2. Classification of In-Plane Optofluidic Lenses

In this part, the in-plane optofluidic lenses will be reported based on their operation mechanisms. Firstly, some example of the refractive lenses based on the fluidic interfaces will be presented. Then the GRIN lenses are discussed. In addition, the in-plane liquid lenses based on other methods are also discussed at the end of this section.

2.1. Interfacial Deformation

The most straightforward method to construct an in-plane optofluidic tunable lens is to use the interfaces between immiscible streams (or liquid-air interface), where the interfacial curvature can be modified by numerous microfluidic techniques [25,26,33]. The general working principle is depicted in Figure 1(b1). In the case of in-plane lens, the geometry modulation can be achieved either by the pressure control or by the hydrodynamic streams. In the pressure-control liquid lens, the curvature of the liquid-air interface is modified by external pumping [25,34]. Tony Huang's group proposed a reconfigurable in-plane liquid lens using fluidic pressure to tune the liquid/air interface in a microfluidic chip [25]. As shown in Figure 2a, this microlens consists of a reconfigurable divergent liquid-air interface and a static polydimethylsiloxane (PDMS) lens. The liquid flows through a straight channel and traps the air in the chamber, forming a liquid-air interface. By adjusting the flow rate, it changes the pressure inside the channel as well as the interfacial radius. It demonstrated the continuous modulation of the focal length by tuning the flow velocity. Behind the lens, there is a chamber for experimental raytracing. Another pressure-controllable liquid-air in-plane lens is demonstrated by Dong et al. [34]. By precisely locating a liquid droplet at the T-shape junction, a tunable in-plane liquid lens is formed in the microchannel (see Figure 2b). This microlens has a tunable focal length from a few hundreds of micrometers to infinite. It can be pneumatically repositioned and removed inside the predefined microchannel. The geometry of the chamber can also be used as a tunable lens by modifying the shape using the pressure control [35]. The pressure-control liquid-air interface is governed by the Laplace law:

$$\Delta P = 2\gamma\kappa = \gamma\left(\frac{1}{R_1} + \frac{1}{R_2}\right) \tag{1}$$

where γ is the surface tension coefficient between the liquid and air, κ is the mean curvature of the liquid-air interface. R_1 (in horizontal) and R_2 (in vertical) are the principal curvature radii of the interface. The ideal liquid-air interface is spherical in both horizontal and vertical directions. The external pump is used to balance at the pressure drop at the interface [25,34].

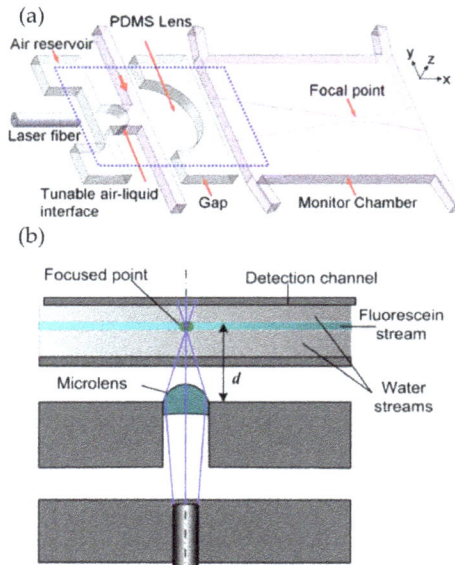

Figure 2. Pressure-control liquid lenses: (**a**) liquid-air interface tuned by flow rate control [25]; (**b**) pneumatically droplet tunable lens [34].

Another way to change the geometry of the in-plane liquid lens is to use the hydrodynamic modulation [26,27,33], in which the fluidic curve is formed and controlled by hydrodynamic force. In this case, two or more immiscible streams (the liquid core and the liquid cladding) are pumped into a specific microchannel to form reconfigurable interfaces. Tuning the ratio of the flow streams enables to continuously modulate the fluidic interfaces. It should be noted that the optical properties of the hydrodynamic stream liquid lens are dependent on the shape of the fluidic chamber. Seow et al. demonstrated a tunable liquid lens by injecting three flow streams into a rectangle-shaped expansion chamber [36], where the liquid with a higher RI acts as the core and the other two streams with a lower RI act as the cladding. Figure 3 shows the schematic designs of the liquid lenses, V_{co} is the flow rate of the core, V_{cll} and V_{clr} are the flow rates of the left and right claddings, respectively. A biconvex lens is formed when $V_{co} > V_{clr} = V_{cll}$, see Figure 3a. And the curvature radius becomes smaller with a higher cladding flow rate. By increasing the value of V_{clr}, the microlens becomes plano-convex (Figure 3b) and then concave-convex (Figure 3c), respectively. Both collimation and focusing have been demonstrated in this type of microlens by tuning the flow rates.

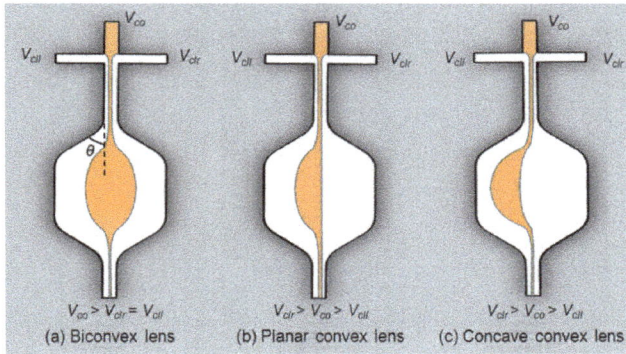

Figure 3. Different curvatures of tunable liquid microlenses via the control of laminar flow rate [36]: (a) biconvex lens; (b) plano-convex lens; (c) concave-convex lens.

Another design of liquid lens uses the circular chamber as shown in Figure 4. Song et al. reported the modeling and experimental results of a tunable lens by injecting three laminar streams into a circular chamber [37]. The liquid-core liquid-cladding lens with perfect curvatures was formed by the circular design. Figure 4 describes the schematic design of the circular liquid lens. In the symmetric state, the lens has a biconvex shape as shown in Figure 4a. By further increasing the flow rate of inlet C, it can tune the lens into the plano-convex shape (Figure 4b) and then the concave-convex shape (Figure 4c). The curvature radius can be tuned from that of the chamber to infinity. As the width of the channel is much smaller than that of the expansion chamber, the model can be approximately described as a source-sink pair model [37]. A reconfigurable biconcave lens was demonstrated by Li et al. [38]. They used the combination of pressure driven flow and electro-osmosis to realize both focusing and diverging in a rectangle chip. Fang et al. proposed a hydrodynamically reconfigurable optofluidic lens, which can be tuned from biconcave to biconvex [27]. Figure 5 depicts the operation principle of the liquid lens. The chamber with two convex ends is used to realize the modulation from biconcave to biconvex. Two immiscible liquids with different RIs are injected into the expansion chamber, where the liquid core (with higher RI) is sandwiched by the liquid claddings (with lower RI). The curvature of the interface is modified by tuning the flow ratio of the core and cladding streams. When the cladding flow is low, the liquid core expands outside into the cladding area, resulting in a biconvex lens (see Figure 5a,b). By increasing the rate of the cladding flow, the curvature of the liquid lens decreases. With the further increase of the cladding flow, the liquid core is compressed into a biconcave shape and the lens becomes negative, as shown in Figure 5c,d. The modulation from biconvex (positive) to biconcave (negative) lens has been demonstrated by adjusting the flow rate. They proposed a two-dimensional quadrupolar flow model to analyze the operation of the liquid lens [27]. As shown in Figure 6, the model has two sources at the left side and two sinks at the right side. The sources and sinks were regarded as dimensionless points. By combining the flow model and the theory of thick lens together, an equation was derived to describe the focal length:

$$f = \frac{-n_1 r^2}{2(n_1 - n_0)[(n_1 - n_0)(s + b) - n_1 r]} \tag{2}$$

where n_0 and n_1 are the RIs of the liquid cladding and liquid core, respectively. And b is half of the distance between the two sources, r is the curvature radius of the liquid interface. The parameter s equals to d or $-d$ when the interface is positive or negative, respectively. By using the combination of a tunable biconvex lens and a reconfigurable liquid prism. Chao et al. demonstrated the controlling of the focal length and the deviation angle of the beam [39].

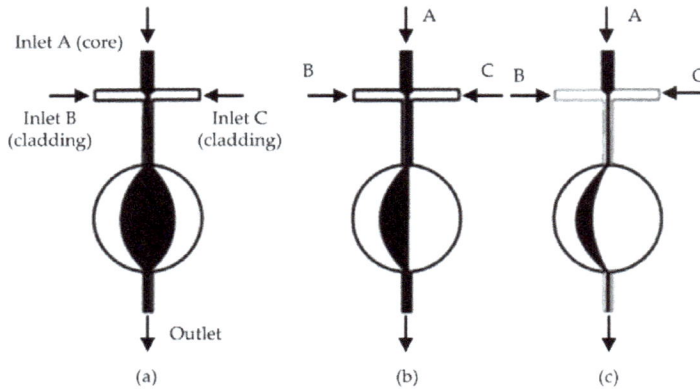

Figure 4. Reconfigurable optofluidic lenses with a circular lens chamber [37]: the lens shape is modified by adjusting the flow rates of the core and cladding streams. (**a**) biconvex lens; (**b**) plano-convex lens; (**c**) concave-convex lens.

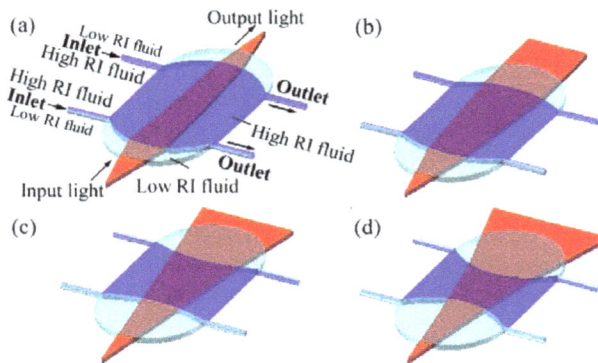

Figure 5. Hydrodynamically reconfigurable optofluidic lens, in which the liquid core (in blue) is sandwiched by the liquid claddings [27]. (**a**) The liquids form a biconvex lens and the beam is focused. (**b**) The beam is collimated when the interface curvature becomes smaller. (**c**,**d**) A biconcave lens is obtained and the beam becomes divergent.

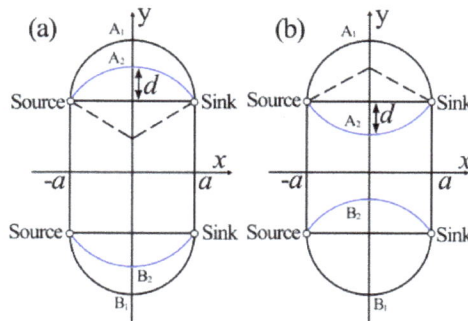

Figure 6. The coordinate of microlens model [27].

2.2. Refractive Index (RI) Modulation

As mentioned above, the RI modulation is another way to alter the optical properties of fluidic components. A simple method to change the RI of the liquid medium is to replace one with another. Seow et al. proposed a tunable planar optofluidic lens using a PDMS lens chamber [40]. By filling the chamber with the mixer of two miscible liquids, the RI was tuned from 1.33 to 1.63. The RI of medium is dependent on several physical properties such as concentration [28,41] and temperature [30]. It can be also changed by external electric field, acoustic field and mechanical strain. The optical propertis of the optofluidic device can be tuned through the modulation of the RI profile, which is also very popular in solid optics. For instance, in a graded index optical fiber, rays follow sinusoidal paths and cross each other periodically. Similarly, rays bend gradually and focus to a focal point, forming the GRIN lens. Compared with the solid materials, the RI modulation of liquid is much easier. By simply changing the concentration of the solutions, the RI change over 0.1 can be achieved [21]. A variety of optofluidic waveguides [19,42,43] and lenses [28,44] base on the diffusion of two miscible solutions have been demonstrated. Temperature gradient is another effective way to form a RI gradient in fluid [30].

A simple method to form a RI gradient within liquid medium is solute diffusion. In a laminar flow inside the microchannel, the concentration gradient is determined by the solution diffusion [45], which can be modulated by the flow rate control. Therefore, a graded RI profile can be achieved using the solution diffusion. Yang et al. proposed an optofluidic RI gradient for lightwave bending and manipulation through the diffusion between ethylene glycol and deionized water [19], in which the RI can be tuned from 1.34 to 1.42. Mao et al. demonstrated a reconfigurable liquid gradient index (L-GRIN) lens with two degrees of freedom using $CaCl_2$ solution as the core and deionized water as the cladding [28]. As shown Figure 7A, the two liquids (the $CaCl_2$ solution and DI water) are injected into the microfluidic chip to establish the gradient profile by diffusion of laminar flows. The rays bend gradually when they meet the RI gradient. Tuning the flow rates of the liquids enables not only to change the focal length, but also to shift the focused beam away from the optical axis, providing another freedom for adaptive optics. Figure 7B depicts the RI distribution along lines 1–5. The RI profile inside the channel follows a hyperbolic secant (HS) function as

$$n^2(x) = n_s^2 + \left(n_0^2 - n_s^2\right)\text{sech}^2(\alpha x) \tag{3}$$

where $n(x)$ is the RI at the given position, n_0 is the RI at the center, n_s is the lowest RI in the liquid medium and α is the gradient parameter. Changing the flow rate enables the modulation of the RI profile as well as the focal length of the lens. Figure 7C shows the RI along line 3 at different flow rates. The ray tracing simulated results in different flow conditions are shown in Figure 7D. It can also shift the focus away from the center using an asymmetric RI profile.

Zhao et al. further improved the performance of the diffusion based optofluidic lens by upgrading the lens design [44], see Figure 8a. By adding a fluidic mixer before the lens section, a HS RI profile can be achieved by precisely controlling the flow rates of the mixer. Borrowed the idea from aberration-free Maxwell's fisheye lens, such a structure is demonstrated to have a lower spherical aberration (see Figure 8b). It is able to focus the beams to different shifted positions on the same focal plane (Figure 8c).

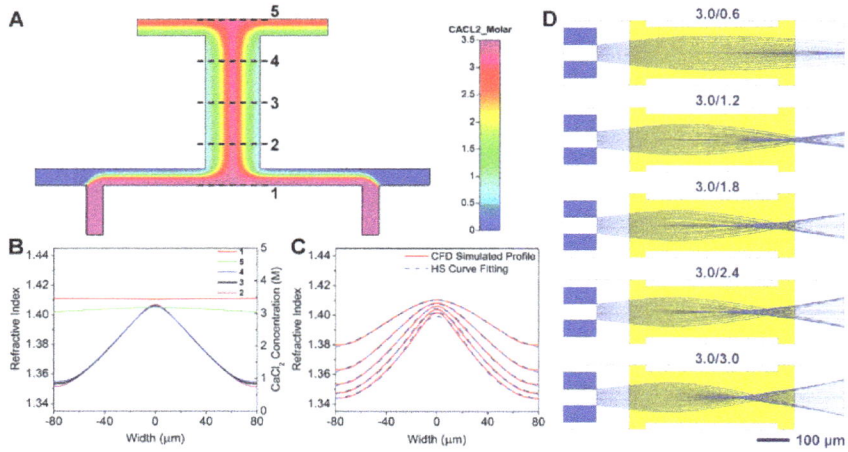

Figure 7. L-GRTN lens with two degrees of freedom [28]. (**A**) Simulated refractive index profile and ray tracing. (**B**) Cross-sectional refractive index distribution at different locations along the flow direction (1, 2, 3, 4 and 5 as indicated in a). (**C**) Refractive index distribution along line 3 (defined in a) at different flow rates. (**D**) Ray tracing results in different flow conditions (3.0/0.6 represents $CaCl_2$ flow rates = 3.0 $\mu L\ m^{-1}$ and H_2O flow rate = 0.6 $\mu L\ m^{-1}$, respectively).

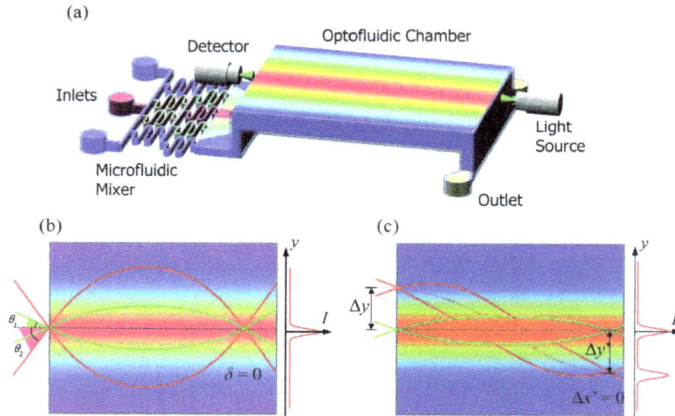

Figure 8. Schematic and working principle of the optofluidic lens [44]. (**a**) Design of the optofluidic chip; (**b**) Spherical aberration; (**c**) Field curvature aberration.

Temperature conduction is another effective way to form a graded RI profile for beam manipulation in microfluidics. According to the thermal lens effect, the RI decreases linearly while the temperature is increased. Therefore, the RI is lower at the hot region. The rays gradually bend while experiencing an inhomogeneous temperature field. As the magnitude of the thermal conduction coefficient is about two orders larger than that of the molecular diffusion coefficient, the thermal lens effect promises a faster response speed. But the thermos-optics coefficient is relative small, which has a value of $1\sim10 \times 10^{-4}\ K^{-1}$. For instance, water has a thermo-optics coefficient of $-1.2 \times 10^{-4}\ K^{-1}$ at 0~80 °C. The thermal-induced RI is at the order of 0.01, which is much smaller as compared to that derived from the concentration gradient. Tang et al. proposed a thermal-induced optical waveguide by the streams at different temperatures [18]. It utilized two streams at higher temperature (the cladding) to sandwich another stream at lower temperature (the core) to form a temperature gradient across the channel. By simply changing the flow rate, the optical properties of the liquid waveguide can be modified. In our previous work, we presented an optofluidic tunable lens using the laser-induced thermal gradient, in which a RI gradient is established in microscale for focusing. As shown in Figure 9, a pump laser is utilized to illuminate the two metal patterns (the yellow pads in Figure 9a), which absorb the light and heat up the flowing liquid (benzyl alcohol, $dn/dT = 4 \times 10^{-4}$ K^{-1}). A temperature induced RI gradient is established inside the microchannel for beam manipulation. Different from the conventional GRIN lenses, this laser-induced thermal lens has a 2D RI gradient, in which the cross-sectional RI follows the square-low parabolic function as described by

$$n(r,z) = n_{c,z}\sqrt{1 - A_z r^2} \qquad (4)$$

where $n(r,z)$ is the RI at point (r,z) and z is the coordinate position along the flow direction. $n_{c,z}$ is the RI at the central position $(r = 0,z)$, and A_z is the parabolic parameter. The simulated 3D- and 2D-RI profiles are shown in Figure 9b,c, respectively. The rays bend gradually and focus to a point while passing the gradient section in between the two metal strips (see Figure 9a). This optofluidic lens allows to use only one liquid. The pump laser enables noncontact modulation and free relocation of the lens region.

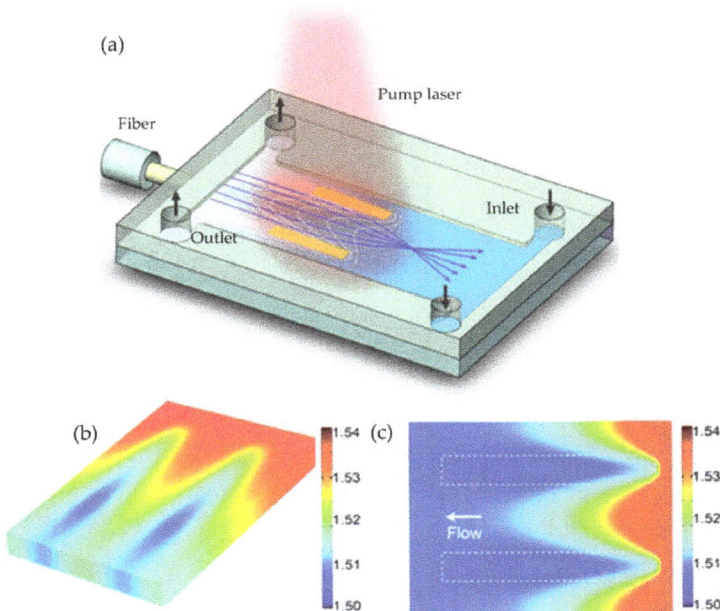

Figure 9. Optofluidic thermal lens using laser-induced thermal gradient [29]: (**a**) schematic design; (**b**) 3D and (**c**) 2D RI profiles.

2.3. Others

Apart from the above mentioned liquid lenses, there are other types of in-plane optofuidic lenses that can also be used for beam manipulation in microfluidic networks.

In the conventional hydrodynamic liquid-liquid lens, isotropic liquids are used as the core and the cladding, which are polarization independent. However, a polarization-dependent device may find special applications in which the polarized light is preferred. A commonly used polarizable liquid is the LC that has a state between isotropic liquid and totally anisotropic solid crystal, resulting in a partially anisotropic fluid. One of the LC phases is nematic, which is usually uniaxial. It has a preferred long axis and a short axis. The nematic LCs have the fluidic properties similar to ordinary organic liquids, but they can be well aligned by a sufficiently strong electric field. Therefore, the nematic LCs have been widely used in electrically reconfigurable optical devices, such as liquid crystal display. Numerous LC lenses have also been demonstrated [46]. The incident light experiences different RIs according to its polarization. The effective RI can be expressed by

$$n_{eff} = n_o n_e \sqrt{\frac{1}{n_o^2 \cos^2 \theta + n_e^2 \sin^2 \theta}} \tag{5}$$

where θ is the angle between the LC rod and the polarization of the incident light, n_e and n_o are the exordinary RI and the ordinary RI of the LC molecule, respectively. It is noted that n_{eff} varies from n_o to n_e according to the wave polarization.

Wee et al. demonstrated an in-plane optofluidic birefringent lens by manipulating the streams of a nematic LC and an isotropic liquid under an external electric field [31]. Figure 10 shows the schematic design, in which the nematic LC and the isotropic liquid are used for the main stream and the surrounding sub-streams, respectively. When an external electric field is applied in the direction perpendicular to the substrate, the LC molecules are reoriented along the electric field, resulting in the birefringent effect in the liquid layer. As n_e and n_o have different values in the nematic LC, the RI difference (i.e., the RI contrast on two sides of the interface) is dependent on the polarization state of the incident light. For the p-polarized light, the effective index equals to n_o, which is smaller than n_e. Therefore, the RI contrast ($\Delta n = n_{eff} - n_i$) of the p-mode is smaller than that of the s-mode. As a consequence, the s-mode has a smaller focal length (see Figure 10b) than that of the p-mode (Figure 10a). This new type of hydrodynamic optofluidic lens can be modulated by either the flow rate or the polarization of the incident light.

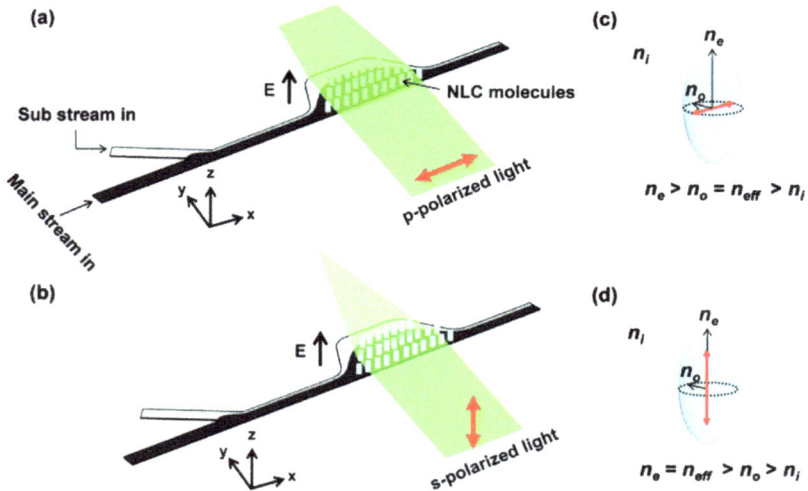

Figure 10. Schematic of the birefringent optofluidic lens [31]: the focusing effect of (**a**) p-polarized and (**b**) s-polarized light. Index ellipsoids are used to describe the RI of (**c**) p-polarized and (**d**) s-polarized light. The red arrows indicate the direction of polarization.

The traditional lenses focus the light due to the reflection or refraction. While in a Fresnel zone plate (FZP), the light is focused by diffraction, which is wavelength dependent. Inspired by this idea, Yang's group demonstrated an optofluidic FZP used a solid-liquid hybrid structure [32] (see Figure 11). It utilizes a microfluidic mixer to prepare the liquid with a specific RI, which is delivered to fill the haft wave zone in the FZP. The radius R_m of the mth half-wave zone is described by

$$R_m = \sqrt{m\lambda f} \tag{6}$$

where λ and f are the wavelength and the focal length, respectively. When m is an integer, a constructive interference appears at the focus. By tuning the RI of the liquid, it demonstrated the real-time modulation of the optical properties, such as peak intensity, spot size and focal length. In addition, this diffractive device is wavelength sensitive, which can selectively focus the desired wavelength. It also enables the switching between focusing, defocusing and collimation by the flow-rate control.

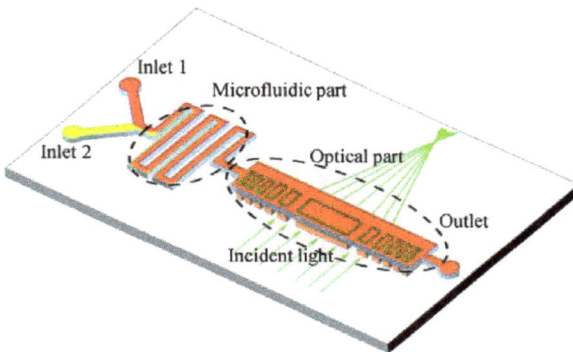

Figure 11. Schematic of the optofluidic Fresnel zone plate (FZP) [32]: it consists of a microfluidic part and an optical part. The former is used to prepare the mixed solution and the latter acts as the tunable FZP.

3. Applications of In-Plane Liquid Lenses

The in-plane optofluidic lenses can be used not only for beam manipulation, but also for some other applications in microfluidic networks, such as particle trapping. Optical tweezers utilize the tightly focused beam to trap particles in the microscale size, providing a nondestructive way to manipulate microparticles. The beam is usually focused by an objective lens with large NA and short focal length. In the traditional optical tweezer, the optical alignment is complicated, making it difficult to integrate it into a miniature lab-on-a-chip system. The emergence of optofluidics makes it easy to manipulate the particle using optical tweezers. In addition, sorting and precise moving of particle in 2D area can be easy achieved in optofluidic networks, which are difficult in conventional optical tweezers.

Yang's group proposed a reconfigurable optofluidic thermal GRIN lens and used it for single cell manipulation in the optofluidic chip [30]. As shown Figure 12, the system consists of two parts: the GRIN lens part and the cell trapping chamber. In the GRIN lens section, benzyl alcohol streams at different temperatures are pumped into the chamber to form a GRIN lens to focus the probe beam. And the living cell is contained in the cell trapping part. The cell is trapped and moved by the focused beam. By regulating the thermal lens, the trapping of the living cell can be modulated at a range of 280 μm. Recently, Aiqun Liu's group has demonstrated the manipulation of particles with sizes of several 10s nanometers using optical potential wells created by an in-plane focused beam and fluidic constraints in optofluidic chips [47,48]. The combination of the quasi-Bessel optical profile and the loosely overdamped potential wells enable the precise manipulation of nanoparticles in the optofluidic channel. They revealed an unprecedentedly meaningful damping scenario that enriches our fundamental understanding of particle kinetics in intriguing optical systems and offered new opportunities for tumor targeting, intracellular imaging, and sorting small particles such as viruses and DNA [48]. The development in particle manipulation using the in-plane beams predicts perspective of the in-plane liquid lens.

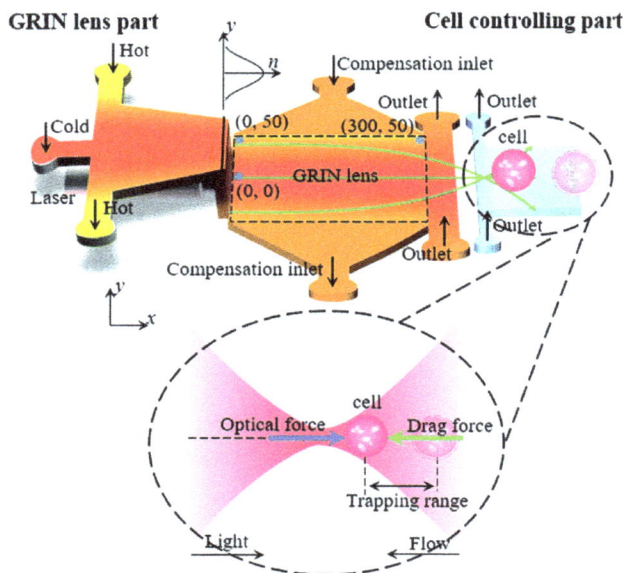

Figure 12. The schematic design of fluidic thermal GRIN lens for cell manipulation [30]: The system includes a lens chamber and a cell trapping chamber. Five streams at different temperatures are injected into the microfluidic chip to form a gradient refractive index across the channel.

The in-plane liquid lens has also been used for particle detection. Flow cytometers have been widely used for particle analysis, sorting and counting. A conventional flow cytometer consists of four main parts, including flow control, light guiding, signal collection and subsequent processing. By the use of optofluidic techniques, the flow cytometer can be integrated into a microfluidic system, reducing the size of the system and making it more portable and durable. The integration allows the benefits of including new optical features on the devices, such as built-in optical alignment, beam shaping, high optical sensitivity and high accuracy [49]. Zhang et al. gave a comprehensive review about the development of optofluidics based flow cytometers [49]. As the detected particles (or cells) are focused to a narrow stream to ensure that they pass through the optical interrogation point one by one, the measured coefficient of variation (CV) of fluorescent beads is strongly dependent on the beam geometry and bead size. As the beam coming out of the waveguide becomes divergent while it strikes at the interrogation region, a 2D lens is used to reshape and focus the beam, enhancing the beam quality for interrogation. In the past, a micro solid lens was fabricated in between the waveguide and the microchannel to improve the performance of the flow cytometer [50]. The precise fabrication of the waveguide and the solid lens system is required to ensure a good focused spot. While a reconfigurable in-plane lens allows the flexible modulation of the beam after fabrication. In addition, the tuning of the spot size makes it possible to get a low CV while detecting particles with different sizes. Goldin et al. proposed an in-plane liquid-filled lens for flow cytometer [51]. Figure 13 shows an optofluidic flow cytometer by Nguyen's group [52]. It utilized a liquid core/liquid cladding lens to focus the light into a microchannel, where the detected particles flow through. A detection fiber was placed at the opposite side of the sample channel to collect the optical signal for subsequent processing. This compact device incorporated the optical elements and flow control part into a single chip to form a flow cytometer, instead of using bulky optics. This optofluidic flow cytometer demonstrated high efficiency and accuracy on particle counting and sizing. The detection of particles with sizes of 5, 10 and 20 μm was demonstrated. A better focused beam can significantly enhance the performance of the flow cytometers. Recently, Yang et al. gave a detail review on the development of the micro-optics for microfluidic analytical applications [53].

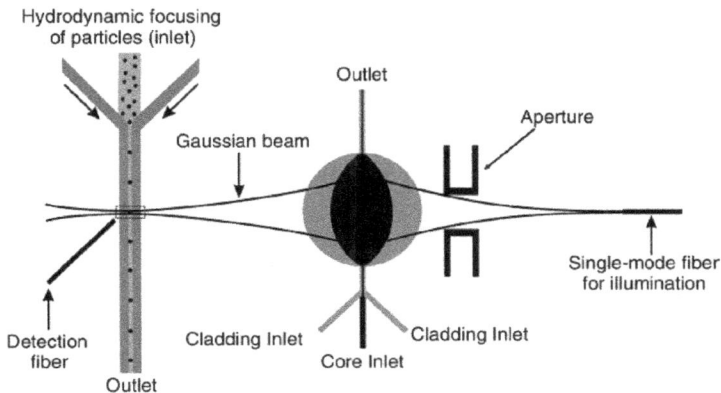

Figure 13. Schematic design of a flow cytometer using optofluidic lens [52].

4. Discussion

Different types of in-plane optofluidic lenses have been demonstrated, providing numerous methods to manipulate the light in a microfluidic network. In the previous refractive lenses, the liquid-liquid (liquid-air) interfaces are spherical, which may lead to longitudinal spherical aberration (the marginal rays are focused closer to the lens than the paraxial rays). Therefore, the beam can not be well focused into the size close to the diffraction limit. The large focused spot size limits the practical applications. For instance, the CVs in the liquid-lens-based (beam waist 23 μm) cytometer [52] are not as good as those conventional cytometers. Howver, the aberration can be well-suppressed in the fluidic GRIN lenses [29,30,44]. For example, the well-designed thermal lens (the spot size 4 μm) has been used to manipulate the living cells in a microfluidic network [30]. As the reflective liquid lens is modulated through pressure control or hydrodynamics, it promises better robustness. While the liquid GRIN lenses usually use the solution diffusion or the temperature gradient, which are susceptible to the ambient fluctuation. In terms of response time, the interfacial deformation takes a few seconds to switch to a new balance state. In comparison, a thermal-induced liquid lens has a faster modulated speed (it is 200 ms in [29]). Another limitation is that most of previous in-plane liquid lenses require continuous supply of liquids, which consumes a lot of liquid and reduces the compatibility of the devices. Although there are still some drawbacks, the in-plane optofluidic lenses find useful applications in lab-on-a-chip systems. In addition, the LC-based and FZP-based liquid lenses enable the polarization and wavelength manipulation in a microfluidic chip. It is foreseeable that more liquid lenses will be incorporated into a chip to achieve versatile microsystems.

5. Conclusions

This article presents a general review of in-plane optofluidic lenses. Based on their working principles and operation methods, they are categorized into three types: refractive lens, gradient index lens and others. The first two cover most of the reported designs, and the last represents some new types, such as birefringent optofluidic lenses and Fresnel zone plate liquid lenses. The in-plane optofluidic lenses have wide tunability, provide a flexible way to manipulate the beam in microfluidic networks and can be easily integrated with lab-on-a-chip systems, making them suitable for transportation and detection of the particle at the same time. It is foreseeable that the in-plane liquid lenses will make the optofluidic networks a versatile platform for research and practical applications. For instance, combing the in-plane beam shaping and microfluidic techniques enables the manipulation and analysis of the particle or cells down to nanometer scale, which may broaden our fundamental understanding of particle kinetics and be useful for biological research. A tightly focused in-plane beam can be used for sample (particle/cell) traping, transportation, separation and detection, making it possible to construct a portable integrated system for biochemical applications.

Acknowledgments: This work is supported by National Natural Science Foundation of China (61377068), Research Grants Council of Hong Kong (N_PolyU505/13, 152184/15E and 152127/17E), and Hong Kong Polytechnic University (G-YBPR, 4-BCAL, 1-ZE14, 1-ZE27 and 1-ZVGH). Acknowledgments should also go to Materials Research Centre, University Research Facility in Material Characterization and Device Fabrication, and University Research Facility in Life Sciences of the Hong Kong Polytechnic University for the technical assistance and facility support.

Author Contributions: Qingming Chen drafted the manuscript, Tenghao Li prepared graphs, Zhaohui Li advised on the technical issues, Jinlin Long helped with the application issues and Xuming Zhang provided the idea and revised the manuscript.

Conflicts of Interest: The authors declare no conflict of interest.

References

1. Psaltis, D.; Quake, S.R.; Yang, C.H. Developing optofluidic technology through the fusion of microfluidics and optics. *Nature* **2006**, *442*, 381–386. [CrossRef] [PubMed]
2. Zhao, Y.; Stratton, Z.S.; Guo, F.; Lapsley, M.I.; Chan, C.Y.; Lin, S.-C.S.; Huang, T.J. Optofluidic imaging: Now and beyond. *Lab Chip* **2013**, *13*, 17–24. [CrossRef] [PubMed]
3. Monat, C.; Domachuk, P.; Eggleton, B.J. Integrated optofluidics: A new river of light. *Nat. Photon.* **2007**, *1*, 106–114. [CrossRef]
4. Fan, X.D.; White, I.M. Optofluidic Microsystems for Chemical and Biological Analysis. *Nat. Photon.* **2011**, *5*, 591–597. [CrossRef] [PubMed]
5. Pang, L.; Chen, H.M.; Freeman, L.M.; Fainman, Y. Optofluidic devices and applications in photonics, sensing and imaging. *Lab Chip* **2012**, *12*, 3543–3551. [CrossRef] [PubMed]
6. Chen, Y.F.; Jiang, L.; Mancuso, M.; Jain, A.; Oncescu, V.; Erickson, D. Optofluidic opportunities in global health, food, water and energy. *Nanoscale* **2012**, *4*, 4839–4857. [CrossRef] [PubMed]
7. Erickson, D.; Sinton, D.; Psaltis, D. Optofluidics for energy applications. *Nat. Photon.* **2011**, *5*, 583–590. [CrossRef]
8. Levy, U.; Shamai, R. Tunable optofluidic devices. *Microfluid. Nanofluidics* **2007**, *4*, 97–105. [CrossRef]
9. Monat, C.; Domachuk, P.; Grillet, C.; Collins, M.; Eggleton, B.J.; Cronin-Golomb, M.; Mutzenich, S.; Mahmud, T.; Rosengarten, G.; Mitchell, A. Optofluidics: A novel generation of reconfigurable and adaptive compact architectures. *Microfluid. Nanofluidics* **2007**, *4*, 81–95. [CrossRef]
10. Erickson, D. Optofluidics. In *Microfluidics Based Microsystems*; Springer: Dordrecht, The Netherlands, 2010; pp. 529–551.
11. Hunt, H.C.; Wilkinson, J.S. Optofluidic integration for microanalysis. *Microfluid. Nanofluidics* **2008**, *4*, 53–79. [CrossRef]
12. Nguyen, N.-T. Micro-optofluidic Lenses: A review. *Biomicrofluidics* **2010**, *4*, 031501. [CrossRef] [PubMed]
13. Fan, X.D.; Yun, S.H. The potential of optofluidic biolasers. *Nat. Methods* **2014**, *11*, 141–147. [CrossRef] [PubMed]

14. Mandal, S.; Erickson, D. Nanoscale optofluidic sensor arrays. *Opt. Express* **2008**, *16*, 1623–1631. [CrossRef] [PubMed]

15. Kühn, S.; Measor, P.; Lunt, E.J.; Phillips, B.S.; Deamer, D.W.; Hawkins, A.R.; Schmidt, H. Loss-based optical trap for on-chip particle analysis. *Lab Chip* **2009**, *9*, 2212–2216. [CrossRef] [PubMed]

16. Schmidt, B.S.; Yang, A.H.; Erickson, D.; Lipson, M. Optofluidic trapping and transport on solid core waveguides within a microfluidic device. *Opt. Express* **2007**, *15*, 14322–14334. [CrossRef] [PubMed]

17. Shopova, S.I.; Zhou, H.Y.; Fan, X.D.; Zhang, P. Optofluidic ring resonator based dye laser. *Appl. Phys. Lett.* **2007**, *90*, 221101. [CrossRef]

18. Tang, S.K.Y.; Mayers, B.T.; Vezenov, D.V.; Whitesides, G.M. Optical waveguiding using thermal gradients across homogeneous liquids in microfluidic channels. *Appl. Phys. Lett.* **2006**, *88*, 061112. [CrossRef]

19. Yang, Y.; Liu, A.Q.; Chin, L.K.; Zhang, X.M.; Tsai, D.P.; Lin, C.L.; Lu, C.; Wang, G.P.; Zheludev, N.I. Optofluidic waveguide as a transformation optics device for lightwave bending and manipulation. *Nat. Commun.* **2012**, *3*, 651. [CrossRef] [PubMed]

20. Kuiper, S.; Hendriks, B.H.W. Variable-focus liquid lens for miniature cameras. *Appl. Phys. Lett.* **2004**, *85*, 1128–1130. [CrossRef]

21. Chiu, C.P.; Chiang, T.J.; Chen, J.K.; Chang, F.C.; Ko, F.-H.; Chu, C.W.; Kuo, S.-W.; Fan, S.K. Liquid Lenses and Driving Mechanisms: A Review. *J. Adhes. Sci. Technol.* **2012**, *26*, 1773–1788. [CrossRef]

22. Mishra, K.; van den Ende, D.; Mugele, F. Recent Developments in Optofluidic Lens Technology. *Micromachines* **2016**, *7*, 102. [CrossRef]

23. Xu, S.; Ren, H.W.; Wu, S.T. Dielectrophoretically tunable optofluidic devices. *J. Phys. D Appl. Phys.* **2013**, *46*, 483001. [CrossRef]

24. Krogmann, F.; Monch, W.; Zappe, H. Electrowetting for Tunable Microoptics. *J. Microelectromech. Syst.* **2008**, *17*, 1501–1512. [CrossRef]

25. Shi, J.; Stratton, Z.; Lin, S.C.S.; Huang, H.; Huang, T.J. Tunable optofluidic microlens through active pressure control of an air–liquid interface. *Microfluid. Nanofluidics* **2009**, *9*, 313–318. [CrossRef]

26. Mao, X.L.; Waldeisen, J.R.; Juluri, B.K.; Huang, T.J. Hydrodynamically tunable optofluidic cylindrical microlens. *Lab Chip* **2007**, *7*, 1303–1308. [CrossRef] [PubMed]

27. Fang, C.L.; Dai, B.; Xu, Q.; Zhuo, R.; Wang, Q.; Wang, X.; Zhang, D.W. Hydrodynamically reconfigurable optofluidic microlens with continuous shape tuning from biconvex to biconcave. *Opt. Express* **2017**, *25*, 888–897. [CrossRef] [PubMed]

28. Mao, X.L.; Lin, S.C.S.; Lapsley, M.I.; Shi, J.; Juluri, B.K.; Huang, T.J. Tunable Liquid Gradient Refractive Index (L-GRIN) lens with two degrees of freedom. *Lab Chip* **2009**, *9*, 2050–2058. [CrossRef] [PubMed]

29. Chen, Q.M.; Jian, A.Q.; Li, Z.H.; Zhang, X.M. Optofluidic tunable lenses using laser-induced thermal gradient. *Lab Chip* **2016**, *16*, 104–111. [CrossRef] [PubMed]

30. Liu, H.L.; Shi, Y.; Liang, L.; Li, L.; Guo, S.S.; Yin, L.; Yang, Y. A liquid thermal gradient refractive index lens and using it to trap single living cell in flowing environments. *Lab Chip* **2017**, *17*, 1280–1286. [CrossRef] [PubMed]

31. Wee, D.; Hwang, S.H.; Song, Y.S.; Youn, J.R. Tunable optofluidic birefringent lens. *Soft Matter* **2016**, *12*, 3868–3876. [CrossRef] [PubMed]

32. Shi, Y.; Zhu, X.Q.; Liang, L.; Yang, Y. Tunable focusing properties using optofluidic Fresnel zone plates. *Lab Chip* **2016**, *16*, 4554–4559. [CrossRef] [PubMed]

33. Tang, S.K.; Stan, C.A.; Whitesides, G.M. Dynamically reconfigurable liquid-core liquid-cladding lens in a microfluidic channel. *Lab Chip* **2008**, *8*, 395–401. [CrossRef] [PubMed]

34. Liang, D.; Jiang, H.R. Selective Formation and Removal of Liquid Microlenses at Predetermined Locations Within Microfluidics through Pneumatic Control. *J. Microelectromech. Syst.* **2008**, *17*, 381–392. [CrossRef]

35. Hsiung, S.K.; Lee, C.H.; Lee, G.B. Microcapillary electrophoresis chips utilizing controllable micro-lens structures and buried optical fibers for on-line optical detection. *Electrophoresis* **2008**, *29*, 1866–1873. [CrossRef] [PubMed]

36. Seow, Y.C.; Liu, A.Q.; Chin, L.K.; Li, X.C.; Huang, H.J.; Cheng, T.H.; Zhou, X.Q. Different curvatures of tunable liquid microlens via the control of laminar flow rate. *Appl. Phys. Lett.* **2008**, *93*, 084101. [CrossRef]

37. Song, C.; Nguyen, N.T.; Tan, S.H.; Asundi, A.K. Modelling and optimization of micro optofluidic lenses. *Lab Chip* **2009**, *9*, 1178–1184. [CrossRef] [PubMed]

38. Li, H.; Song, C.; Luong, T.D.; Nguyen, N.T.; Wong, T.N. An electrokinetically tunable optofluidic bi-concave lens. *Lab Chip* **2012**, *12*, 3680–3687. [CrossRef] [PubMed]

39. Chao, K.S.; Lin, M.S.; Yang, R.J. An in-plane optofluidic microchip for focal point control. *Lab Chip* **2013**, *13*, 3886–3892. [CrossRef] [PubMed]

40. Seow, Y.C.; Lim, S.P.; Lee, H.P. Optofluidic variable-focus lenses for light manipulation. *Lab Chip* **2012**, *12*, 3810–3815. [CrossRef] [PubMed]

41. Yang, Y.; Chin, L.K.; Tsai, J.M.; Tsai, D.P.; Zheludev, N.I.; Liu, A.Q. Transformation optofluidics for large-angle light bending and tuning. *Lab Chip* **2012**, *12*, 3785–3790. [CrossRef] [PubMed]

42. Liu, H.L.; Zhu, X.Q.; Liang, L.; Zhang, X.M.; Yang, Y. Tunable transformation optical waveguide bends in liquid. *Optica* **2017**, *4*, 839–846. [CrossRef]

43. Shi, Y.; Liang, L.; Zhu, X.Q.; Zhang, X.M.; Yang, Y. Tunable self-imaging effect using hybrid optofluidic waveguides. *Lab Chip* **2015**, *15*, 4398–4403. [CrossRef] [PubMed]

44. Zhao, H.T.; Yang, Y.; Chin, L.K.; Chen, H.F.; Zhu, W.M.; Zhang, J.B.; Yap, P.H.; Liedberg, B.; Wang, K.; Wang, G.; et al. Optofluidic lens with low spherical and low field curvature aberrations. *Lab Chip* **2016**, *16*, 1617–1624. [CrossRef] [PubMed]

45. Wolfe, D.B.; Vezenov, D.V.; Mayers, B.T.; Whitesides, G.M.; Conroy, R.S.; Prentiss, M.G. Diffusion-controlled optical elements for optofluidics. *Appl. Phys. Lett.* **2005**, *87*, 181105. [CrossRef]

46. Lin, H.C.; Chen, M.S.; Lin, Y.H. A Review of Electrically Tunable Focusing Liquid Crystal Lenses. *Trans. Electr. Electron. Mater.* **2011**, *12*, 234–240. [CrossRef]

47. Shi, Y.Z.; Xiong, S.; Chin, L.K.; Yang, Y.; Zhang, J.B.; Ser, W.; Wu, J.H.; Chen, T.N.; Yang, Z.C.; Hao, Y.L.; et al. High-resolution and multi-range particle separation by microscopic vibration in an optofluidic chip. *Lab Chip* **2017**, *17*, 2443–2450. [CrossRef] [PubMed]

48. Shi, Y.Z.; Xiong, S.; Chin, L.K.; Zhang, J.B.; Ser, W.; Wu, J.H.; Chen, T.N.; Yang, Z.C.; Hao, Y.L.; Liedberg, B.; et al. Nanometer-precision linear sorting with synchronized optofluidic dual barriers. *Sci. Adv.* **2018**, *4*, eaao0773. [CrossRef] [PubMed]

49. Zhang, Y.S.; Watts, B.R.; Guo, T.Y.; Zhang, Z.Y.; Xu, C.Q.; Fang, Q.Y. Optofluidic Device Based Microflow Cytometers for Particle/Cell Detection: A Review. *Micromachines* **2016**, *7*, 70. [CrossRef]

50. Watts, B.R.; Zhang, Z.Y.; Xu, C.Q.; Cao, X.D.; Lin, M. Scattering detection using a photonic-microfluidic integrated device with on-chip collection capabilities. *Electrophoresis* **2014**, *35*, 271–281. [CrossRef] [PubMed]

51. Godin, J.; Lien, V.; Lo, Y.H. Demonstration of two-dimensional fluidic lens for integration into microfluidic flow cytometers. *Appl. Phys. Lett.* **2006**, *89*, 061106. [CrossRef]

52. Song, C.; Luong, T.D.; Kong, T.F.; Nguyen, N.T.; Asundi, A.K. Disposable flow cytometer with high efficiency in particle counting and sizing using an optofluidic lens. *Opt. Lett.* **2011**, *36*, 657–659. [CrossRef] [PubMed]

53. Yang, H.; Gijs, M.A.M. Micro-optics for microfluidic analytical applications. *Chem. Soc. Rev.* **2018**, in press. [CrossRef] [PubMed]

micromachines

MDPI

Review

Optofluidics Refractometers

Cheng Li [1], Gang Bai [2,3], Yunxiao Zhang [2], Min Zhang [1] and Aoqun Jian [2,3,*]

[1] State Key Laboratory of Information Photonics and Optical Communications, Beijing University of Posts and Telecommunications, No. 10, Xitucheng Road, Haidian District, Beijing 100876, China; lichn@bupt.edu.cn (C.L.); mzhang@bupt.edu.cn (M.Z.)

[2] MicroNano System Research Center, College of Information and Computer Science, Taiyuan University of Technology, Taiyuan 030024, China; baigang0308@link.tyut.edu.cn (G.B.); zhangyunxiao1895@link.tyut.edu.cn (Y.Z.)

[3] Key Laboratory of Advanced Transducers and Intelligent Control System, Shanxi Province and Ministry of Education, Taiyuan 030024, China

* Correspondence: jianaoqun@tyut.edu.cn; Tel.: +86-0351-601-0029

Received: 31 January 2018; Accepted: 16 March 2018; Published: 20 March 2018

Abstract: Refractometry is a classic analytical method in analytical chemistry and biosensing. By integrating advanced micro- and nano-optical systems with well-developed microfluidics technology, optofluidics are shown to be a powerful, smart and universal platform for refractive index sensing applications. This paper reviews recent work on optofluidic refractometers based on different sensing mechanisms and structures (e.g., photonic crystal/photonic crystal fibers, waveguides, whisper gallery modes and surface plasmon resonance), and traces the performance enhancement due to the synergistic integration of optics and microfluidics. A brief discussion of future trends in optofluidic refractometers, namely volume sensing and resolution enhancement, are also offered.

Keywords: refractometry; refractive index; microfluidics

1. Introduction

Refractive index (RI), a basic physical substance property, can be used to measure solute concentration and purity in transparent liquor such as Salinity and Brix. RI is highly sensitive and precise, enabling it to monitor extreme variations in tiny particles in solution, which can be used to quantitatively analyze chemical components. For example, 10^{-9} RIU is equivalent to 1 femto mol/L of salt in water. Compared to other measuring methods, RI measurement does not actually affect the properties of the analyst and offers real-time, convenient analysis of liquid composition (e.g., label-free analysis of various bio-samples, including DNA and protein). Therefore, refractometry with ultra-high sensitivity has great potential in environmental protection [1–3], drinking water safety [4,5] and biomedical applications [6].

Microfluidics has achieved great progress recently due to its own excellent performance in fluidic handling [7–9], micro-environment control [10–12] and signal amplification [13,14]. By integrating advanced micro- and nano-optical systems with well-developed microfluidics technology, optofluidics has ushered in a new era of lab-on-a-chip functionality [15–22], including biochemical sensing with optical measurement [23], optofluidic imaging [24], and light-driven manipulation [25,26]. In the case of RI sensing, many valuable review papers such as Fan's [27] have found that synergistic integration creates unique characteristics that promote the performance and function of biological/chemical analysis. First, in some well-designed structures, the analyte can be selectively delivered to a location with maximum light-analyte interaction, which can significantly enhance sensors' sensitivity and resolution. Second, an extremely small analyte volume (i.e., nL) and some related treatments of biological samples, such as cultivating, sorting, trapping, and purification, can be achieved with microfluidics technology. Other analytical methods, including chromatography, electrophoresis, and Raman scattering, can be cascaded with the RI sensing function to carry out complex analysis. Third, issues related to integration, namely alignment and packaging, can be easily solved, and the device's volume can be reduced substantially. Moerover, device commercialization will facilitate the development of portable, cost-effective, and highly sensitive bio/chemical analysis instruments.

In this paper, recent work pertaining to optofluidic refractometers, based on different sensing mechanisms and structures, is sequentially reviewed. Particularly, this review focuses on the brilliant designs that the optical sensing structures, based on their own various characteristics, synergistically integrated with microfluidics to optimize their sensing performances. These designs enhance the resolutions of the sensors, expand the analysis functions of the sensors or solve issues of sensor applications. The revolution tracks of this integration, showing the elegance of the device design, are also presented. A discussion regarding the field's ongoing development is also offered with the hope to inspire more new ideas from the readers.

2. RI Sensing and Technologies

2.1. Photonic Crystal Fibers

As a classic photonic structure, photonic crystal fibers (PCFs) appear to be an ideal platform for the realization of novel optofluidic sensors [28]. P. Domachuk et al. proposed a compact refractometer utilizing a Fabry–Pérot cavity (FPC) etalon formed between two aligned Bragg grating fibers (one-dimensional PCFs), which located on either side of the microfluidic channel to contain the fluid to be tested [15]. The resolution of the RI sensor based on PCFs can be further improved if the microstructured architecture in the PCFs is filled by analyte [29]. Its waveguide nature ensures strong light-analyte interaction along all PCFs. Some smart structures are designed to feed the analyte into the PCFs' hollow gap. Many research efforts contribute to design the deliberate structures to achieve the convenient load of liquid sample, and employ special modes to improve the sensitivity of the sensors. For example, C. Wu et al. presented the fabrication and characterization of an in-line photonic crystal fiber microfluidic refractometer outfitted with a C-shaped fiber [30] (Figure 1a). The C-shaped fiber, placed between the PCF and single-mode fiber, achieved two functions simultaneously: in-line optical signal coupling and analyte fluid feeding. Using an arc discharge pre-treatment technique, small air hole voids near the surface were sealed, so only the two central large air hole channels were employed for RI sensing; thus, device sensitivity increased by 70% due to higher power density. Similarly, N. Zhang et al. utilized a side-channel photonic crystal fiber with side-polished single mode fibers to form optofluidic microchannels [31] (Figure 1b). A long-period grating combined with intermodal interference between LP_{01} and LP_{11} core modes was used to sense the liquid's RI in the side channel.

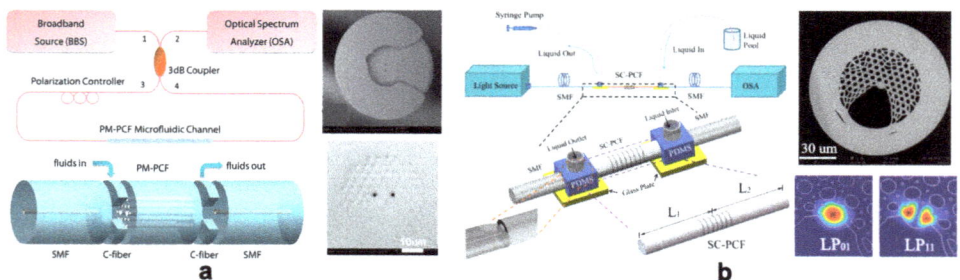

Figure 1. (**a**) Scheme and transverse section graph of the SMF-C-PCF-C-SMF microfluidic device. Reprinted with permission from [30]. Copyright (2014) RSC; (**b**) Scheme of the in-line optofluidic sensing platform and SEM image of the SC-PCF and Simulated intensity distribution of LP_{01} mode and LP_{11} mode at the wavelength of 1550 nm. Reprinted with permission from [31]. Copyright (2016) OSA.

2.2. Planar Optical Waveguides

An integrated planar optical waveguide (POW) has been utilized for RI sensing for several decades [32]. In this type of sensor, light propagates along the solid waveguide, around which an evanescent field interacts with the analyte to induce phase shift or intensity variation. The light-analyte interaction is naturally limited by the evanescent field's range per unit length. Therefore, confining light in the fluidic waveguide is an effective way to achieve strong light-analyte interaction. The liquid core anti-resonant reflecting optical waveguide (ARROW) is a novel photonic structure with tightly integrated optical and fluidic structures [33] (Figure 2). Particularly, Campopiano et al. firstly demonstrated a bulk refractometer based on multimode liquid ARROW structure [34] (Figure 3a,b). And G. Testa et al. offered a comprehensive review of an ARROW-based device's operation principles and applications [35]. Furthermore, the RI microsensor's sensitivity can be improved by using the ARROW waveguide as interferometer arms [36] (Figure 3c) and part of the ring resonators [37] (Figure 3d). In addition, the waveguide can act as a carrier, on which 2D materials can be deposited, for novel RI sensors developments. Surface plasmon waves or evanescent wave are tuned by the liquid medium on the surface of Graphene [38] and MoS_2 [39] to realize RI sensing.

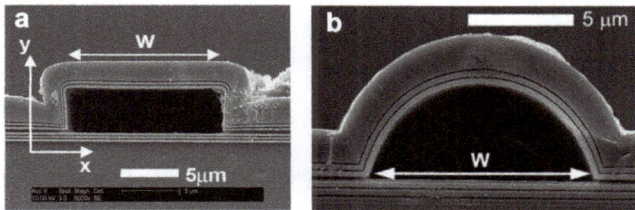

Figure 2. Scanning electron microscope (SEM) images of hollow-core anti-resonant reflecting optical waveguides (ARROWs) with (**a**) rectangular and (**b**) arch-shaped cross sections fabricated by surface micromachining process. Reprinted with permission from [33]. Copyright (2005) OSA.

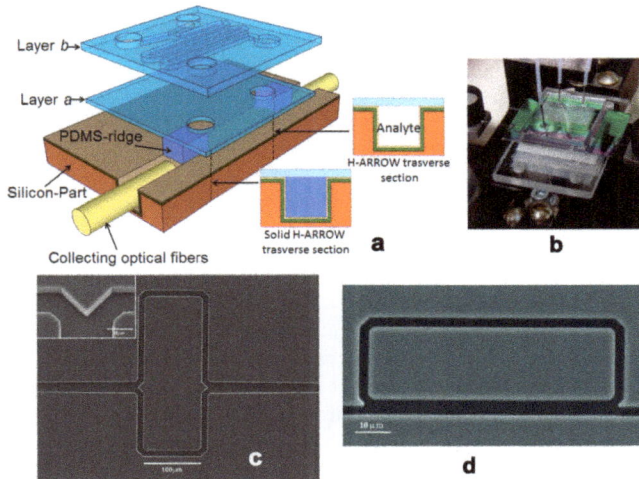

Figure 3. Schematic (**a**) and fabricated chip (**b**) of the sensor based on hybrid ARROW optofluidic platform. Reprinted with permission from [34]. Copyright (2014) OSA; (**c**) SEM picture of liquid-core ARROW. Reprinted with permission from [36]. Copyright (2010) OSA; (**d**) SEM picture of the integrated silicon optofluidic ring resonator. Reprinted with permission from [37]. Copyright (2010) AIP.

2.3. Whisper Gallery Mode

Since F. Vollmer piloted the application of whisper gallery mode (WGM) in protein detection, continued efforts have explored the detection resolution's potential [40]. A series of well-designed microtoroid resonators were fabricated by L. Yang and Y.-F. Xiao, and related intelligent sensing schemes and noise control methods have also been proposed and optimized [41]. Due to the resonators' ultra-high Q values, the microsensors have an extremely low detection limit; single nanoparticles have been detected successfully. However, when applying a typical cavity-taper coupling system in an optofluidic system, uncontrollable analyte flow in the microchip may exert a negative effect on cavity-taper coupling, diminishing the Q value to some extent. Thin-wall cylindrical capillaries with spatial analyte-taper separation comprise an alternative scheme that is more convenient for optofluidic integration [42,43] (Figure 4a–d). Since high-Q WGMs in deformed microcavities can be excited by free space coupling [44] (Figure 4e), it is a feasible way to achieve high cavity coupling efficiency in microfluidic chips.

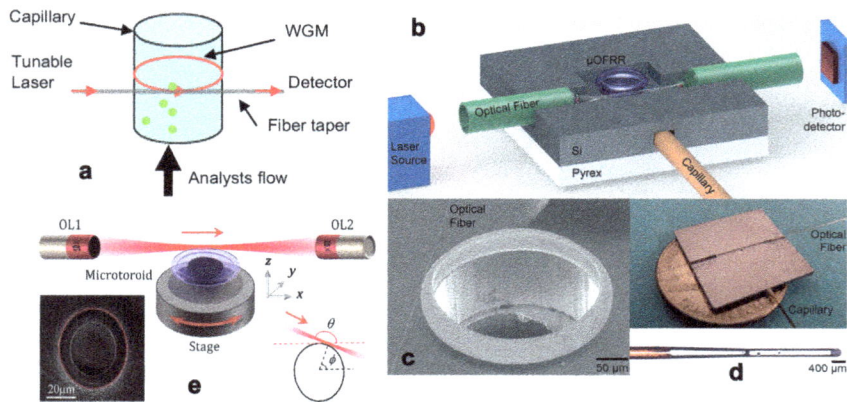

Figure 4. (**a**) The optofluidic ring resonator based on thin-walled capillary. Reprinted with permission from [42]. Copyright (2014) Elsevier; Schematic diagram (**b**), SEM image (**c**) and photograph (**d**) of the μOFRR sensor. Reprinted with permission from [43]. Copyright (2014) RSC; (**e**) Schematic diagram of the experimental setup for free-space coupling between a laser beam and a deformed toroidal microcavity. Reprinted with permission from [44]. Copyright (2016) OSA.

2.4. Surface Plasmon Resonance and Localized Surface Plasmon Resonance

Surface plasmon resonance/localized surface plasmon resonance (SPR/LSPR) refer to the excitation of collective electron charge oscillations on planar metal surfaces or onto the surface of metallic nanoparticles by incident light. The oscillation leads to a wavelength-dependent reduction in the overall reflection, which presents an absorption peak in the reflection/scatter spectrum. As the SPR is very sensitive to the RI near the metal surface (within 300 nm), it is widely used to develop RI sensors [45] and biosensors [46]. Unlike most other types of sensing schemes, metal plane/particle/nanohole arrays represent an indispensable part of the sensing system, which also offers unique opportunities for function appending and upgrading. A. Barik et al. utilized a gold nanohole array-based optofluidic device for label-free detection of analyte molecules, of which the nanohole array also generated gradient dielectrophoretic force to accumulate the measured biological analytes [47] (Figure 5a,b). Real-time detection was over 1000 times faster than the classic diffusion method for 1 pM analyte concentration. S. Kang et al. combined RI sensing with surface-enhanced Raman spectroscopy (SERS) in silver–gold layered bimetallic plasmonic crystals to conduct quantitative and qualitative measurements simultaneously [48]. D. Zhang et al. developed a unique nanoscale cup array (nanoCA) coupling electrochemistry to LSPR spectroscopy measurement, offering a novel method by which to evaluate complex electrochemical reaction processes [49] (Figure 5c–e).

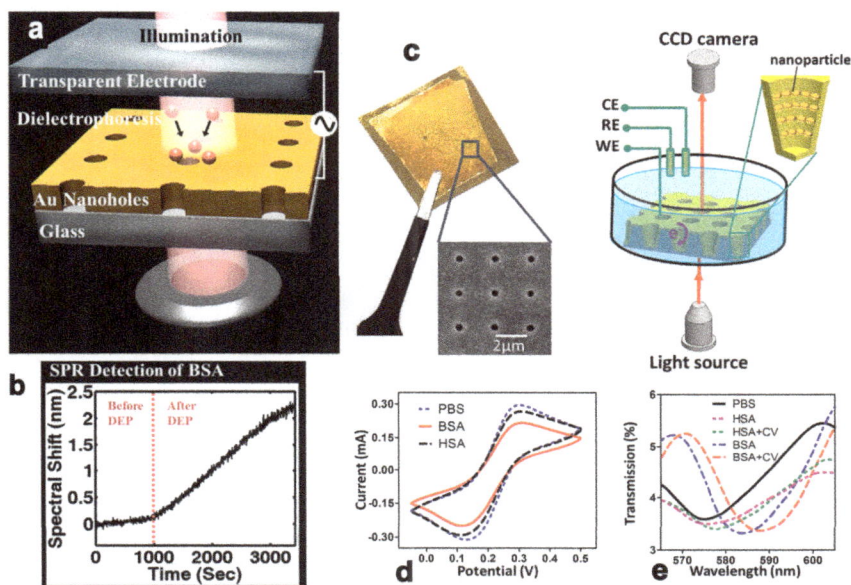

Figure 5. Schematic of the experimental setup for dielectrophoretic concentration of analyte molecules (**a**) and spectral shift of BSA's SPR detection (**b**). Reprinted with permission from [47]. Copyright (2014) ACS; The nanoCA for electrochemical and LSPR measurement (**c**) and Transmission spectrum (**d**) of nanoCA with PBS, 100 μg/mL HSA and HSA plus CV scanning, 100 μg/mL BSA and BSA plus synchronous CV scanning and statistic (**e**) for shifts in dip wavelength of HSA, HSA plus CV, BSA and BSA plus CV. Reprinted with permission from [49]. Copyright (2015) Elsevier.

3. Discussion and Outlook

3.1. Parameters for Sensor Characterization

Sensitivity, defined as the magnitude in shift of the characteristic wavelength versus the change in a sample's RI, is a key parameter describing RI sensor performance. However, for measurement response down to the sensor's detection limit (i.e., resolution), many factors should be considered: the shape of the resonant peak, noise sources, and signal intensity, among others. According to X. D. Fan's detailed analysis, a sensor with lower sensitivity but a sharper resonance peak (i.e., a higher-quality factor) has higher resolution [50]. Thus, to accurately describe sensor performance, a comprehensive evaluation that includes sensitivity and quality factors is crucial. Several novel parameters, such as figure of merit (FOM) [51] or detectivity [52], have been proposed in some papers. FOM is defined as the ratio of sensitivity to full wave at half maximum (FWHM) as Equation (1), which can take both of the sensitivity and Q value into consideration, thus the performance of the sensor can be characterized accurately by using a single parameter.

$$\text{FOM} = \frac{\text{Sensitivity}}{\text{FWHM}} \tag{1}$$

Table 1 summarized the critical parameters for qualifying the performances of the RI sensors based on different principles. Although the FOM value (related to resolution/detection limit) is a key parameter for the RI sensors, some other characteristics (detection range, sample volume, cost-effective ratio, portability and capability of integration with other analyzing instruments) are of importance and should be taken into consideration in the specific biochemical applications.

Table 1. Critical parameters of the refractive index (RI) sensors based on different principles.

Working Principle	Sensitivity	Q Factor	FOM	Detection Limit	Analyte	Reference
PCF	8699 nm/RIU	-	-	4.0×10^{-6} RIU	-	[30]
PCF	1145 nm/RIU	-	-	-	-	[31]
POW	260 nm/RIU	800	-	-	-	[37]
POW	1920 nm/RIU	-	-	5.2×10^{-7} RIU	-	[38]
WGM	1.84 pm/mM	4×10^5	-	0.035 mM	Glucose	[42]
WGM	0.018 pm/mg m^{-3}	11,500	-	6.9 ppm	Benzene	[43]
SPR	-	-	-	1 pM	BSA	[47]
SPR	~104 nm/RIU	-	-	-	-	[49]
FPC	960 nm/RIU	600	18.79	0.01 RIU	-	[53]
FPC	907 nm/RIU	400	9	1.7×10^{-5} RIU	-	[54]

3.2. New Areas for Exploration: Volume Sensing

Currently, most RI sensors are developed based on near-field optics, which uses evanescent waves in the subwavelength region. Dramatically decayed evanescence has marked light-analyte interaction, indicating high sensitivity, but its spatial interaction is intrinsically limited by the attenuation characteristics of evanescent waves. On the other hand, some bio-analytes (e.g., eukaryotic cells) tend to measure 20–30 μm in diameter; in this case, organisms deep inside the cells are outside of the evanescent field and cannot be measured accurately. For RI sensors working on evanescent waves, only the sample in contact with the sensing surface can be measured, which hinders the detection of naturally suspended samples. Furthermore, because the spatial measurement range is expanded from nearly 2D to 3D, volume sensing is a feasible way to increase light-analyte interaction. In volume sensing, the lightwave can completely permeate the targeted sample and detect every particle in the solution—not only the sample attached to the sensing surface. Hence, this method is particularly useful for monolithic biological samples (eukaryotic cells and tissues) and analytes in low-concentration solutions, which is a rising research subject offering clear advantages in a host of applications.

Fabry–Pérot (FP) etalon is considered to be a promising choice for the volume sensing application due to its simple structure and growing performance [16]. Some recent research demonstrated that open-access optical cavities with high Q factor and low mode volume can be achieved by utilizing micro-scale curved-mirrors [55], or even a spherical mirror on a fiber tip-end and an assorted planar mirror [56]. Since RI of living kidney cells [22,57,58], these high performance FP RI sensors are inspired to have bright future in studying cell physiology and pathology.

3.3. Advanced Methods for Performance Enhancement

According to Equation (1), FOM can be optimized from two angles: sensitivity and quality factor. Improving sensitivity depends mainly on enhancing the light-analyte interaction; however, strong light-analyte interaction also indicates intense absorption (i.e., a low Q factor) from solvent, where are often water or phosphate-buffered solution with certain concentration. Therefore, spatially accumulating or attracting interested particles to the area with the strongest light-analyte interaction via external field/force (dielectrophoretic [47], ultrasonic, and magnetic methods [59]) or a microfluidic sorting structure is an effective way to avoid the influence of solvent absorption. Furthermore, cascaded/hybrid structures, such as coupled resonator-induced transparency (CRIT) [60–62] and Vernier effect [26], improve the Q factor and sensitivity synchronously and contribute to resolving the interaction-absorption dilemma.

Micromachines **2018**, *9*, 136

4. Conclusions

This review article has summarized the prominent designs for RI sensing using optofluidic technology. Excellent performance of intelligent designs mentioned in paper indicates a prosperous prospect of optofluidics in bio/chemical analysis. The synergy of photonics and microfluidics offers a great opportunity to achieve the device's performance improvements and functional extension. And microfluidics technology facilitates devices with portable and cost-effective features, providing steady motivation for their commercialization promotion. In future, more and more subtle and powerful devices will be developed to meet the growing need of measurement in biomedical applications.

Acknowledgments: This study was financially supported by the National Natural Science Foundation of China (No. 61501316), the Shanxi Provincial Foundation for Returned Scholars (2015-047), 863 project (2015AA042601), Fund of State Key Laboratory of Information Photonics and Optical Communications (Beijing University of Posts and Telecommunications).

Author Contributions: Cheng Li wrote the whole manuscript. Gang Bai reviewed the manuscript. Yunxiao Zhang and Min Zhang participated in the discussion. Aoqun Jian supervised the work.

Conflicts of Interest: The authors declare no conflict of interest.

References

1. Rotiroti, L.; De Stefano, L.; Rendina, I.; Moretti, L.; Rossi, A.M.; Piccolo, A. Optical microsensors for pesticides identification based on porous silicon technology. *Biosens. Bioelectron.* **2005**, *20*, 2136–2139. [CrossRef] [PubMed]
2. Park, J.-H.; Byun, J.-Y.; Yim, S.-Y.; Kim, M.-G. A localized surface plasmon resonance (LSPR)-based, simple, receptor-free and regeneratable Hg^{2+} detection system. *J. Hazard. Mater.* **2016**, *307*, 137–144. [CrossRef] [PubMed]
3. Bao, B.; Melo, L.; Davies, B.; Fadaei, H.; Sinton, D.; Wild, P. Detecting supercritical CO_2 in brine at sequestration pressure with an optical fiber sensor. *Environ. Sci. Technol.* **2012**, *47*, 306–313. [CrossRef] [PubMed]
4. Yu, J.; Huang, W.; Chin, L.; Lei, L.; Lin, Z.; Ser, W.; Chen, H.; Ayi, T.; Yap, P.; Chen, C. Droplet optofluidic imaging for λ-bacteriophage detection via co-culture with host cell escherichia coli. *Lab Chip* **2014**, *14*, 3519–3524. [CrossRef] [PubMed]
5. Liu, P.; Chin, L.; Ser, W.; Ayi, T.; Yap, P.; Bourouina, T.; Leprince-Wang, Y. An optofluidic imaging system to measure the biophysical signature of single waterborne bacteria. *Lab Chip* **2014**, *14*, 4237–4243. [CrossRef] [PubMed]
6. Liu, P.Y.; Chin, L.K.; Ser, W.; Chen, H.F.; Hsieh, C.M.; Lee, C.H.; Sung, K.B.; Ayi, T.C.; Yap, P.H.; Liedberg, B.; et al. Cell refractive index for cell biology and disease diagnosis: Past, present and future. *Lab Chip* **2016**, *16*, 634–644. [CrossRef] [PubMed]
7. Kaler, K.V.; Prakash, R. Droplet microfluidics for chip-based diagnostics. *Sensors* **2014**, *14*, 23283–23306. [CrossRef] [PubMed]
8. Strohmeier, O.; Keller, M.; Schwemmer, F.; Zehnle, S.; Mark, D.; von Stetten, F.; Zengerle, R.; Paust, N. Centrifugal microfluidic platforms: Advanced unit operations and applications. *Chem. Soc. Rev.* **2015**, *44*, 6187–6229. [CrossRef] [PubMed]
9. Fair, R.B. Digital microfluidics: Is a true lab-on-a-chip possible? *Microfluid. Nanofluid.* **2007**, *3*, 245–281. [CrossRef]
10. Yang, W.; Luo, C.; Lai, L.; Ouyang, Q. A novel microfluidic platform for studying mammalian cell chemotaxis in different oxygen environments under zero-flow conditions. *Biomicrofluidics* **2015**, *9*, 044121. [CrossRef] [PubMed]
11. Chen, Y.-C.; Zhang, Z.; Fouladdel, S.; Deol, Y.; Ingram, P.N.; McDermott, S.P.; Azizi, E.; Wicha, M.S.; Yoon, E. Single cell dual adherent-suspension co-culture micro-environment for studying tumor–stromal interactions with functionally selected cancer stem-like cells. *Lab Chip* **2016**, *16*, 2935–2945. [CrossRef] [PubMed]

12. Chen, Y.-C.; Cheng, Y.-H.; Kim, H.S.; Ingram, P.N.; Nor, J.E.; Yoon, E. Paired single cell co-culture microenvironments isolated by two-phase flow with continuous nutrient renewal. *Lab Chip* **2014**, *14*, 2941–2947. [CrossRef] [PubMed]

13. Chang, C.M.; Chang, W.H.; Wang, C.H.; Wang, J.H.; Mai, J.D.; Lee, G.B. Nucleic acid amplification using microfluidic systems. *Lab Chip* **2013**, *13*, 1225–1242. [CrossRef] [PubMed]

14. Lin, X.; Sun, X.; Luo, S.; Liu, B.; Yang, C. Development of DNA-based signal amplification and microfluidic technology for protein assay: A review. *TrAC Trends Anal. Chem.* **2016**, *80*, 132–148. [CrossRef]

15. Domachuk, P.; Littler, I.C.M.; Cronin-Golomb, M.; Eggleton, B.J. Compact resonant integrated microfluidic refractometer. *Appl. Phys. Lett.* **2006**, *88*, 093513. [CrossRef]

16. Bitarafan, M.H.; DeCorby, R.G. On-chip high-finesse fabry-perot microcavities for optical sensing and quantum information. *Sensors* **2017**, *17*, 1748. [CrossRef] [PubMed]

17. Trichet, A.A.; Foster, J.; Omori, N.E.; James, D.; Dolan, P.R.; Hughes, G.M.; Vallance, C.; Smith, J.M. Open-access optical microcavities for lab-on-a-chip refractive index sensing. *Lab Chip* **2014**, *14*, 4244–4249. [CrossRef] [PubMed]

18. Pang, L.; Chen, H.M.; Freeman, L.M.; Fainman, Y. Optofluidic devices and applications in photonics, sensing and imaging. *Lab Chip* **2012**, *12*, 3543–3551. [CrossRef] [PubMed]

19. Erickson, D.; Sinton, D.; Psaltis, D. Optofluidics for energy applications. *Nat. Photonics* **2011**, *5*, 583–590. [CrossRef]

20. Schmidt, H.; Hawkins, A.R. The photonic integration of non-solid media using optofluidics. *Nat. Photonics* **2011**, *5*, 598–604. [CrossRef]

21. Monat, C.; Domachuk, P.; Eggleton, B. Integrated optofluidics: A new river of light. *Nat. Photonics* **2007**, *1*, 106–114. [CrossRef]

22. Song, W.; Zhang, X.; Liu, A.; Lim, C.; Yap, P.; Hosseini, H.M.M. Refractive index measurement of single living cells using on-chip fabry-pérot cavity. *Appl. Phys. Lett.* **2006**, *89*, 203901. [CrossRef]

23. Bates, K.E.; Lu, H. Optics-integrated microfluidic platforms for biomolecular analyses. *Biophys. J.* **2016**, *110*, 1684–1697. [CrossRef] [PubMed]

24. Zhao, Y.; Stratton, Z.S.; Guo, F.; Lapsley, M.I.; Chan, C.Y.; Lin, S.C.; Huang, T.J. Optofluidic imaging: Now and beyond. *Lab Chip* **2013**, *13*, 17–24. [CrossRef] [PubMed]

25. Seow, Y.C.; Lim, S.P.; Lee, H.P. Optofluidic variable-focus lenses for light manipulation. *Lab Chip* **2012**, *12*, 3810–3815. [CrossRef] [PubMed]

26. La Notte, M.; Troia, B.; Muciaccia, T.; Campanella, C.E.; De Leonardis, F.; Passaro, V.M. Recent advances in gas and chemical detection by vernier effect-based photonic sensors. *Sensors* **2014**, *14*, 4831–4855. [CrossRef] [PubMed]

27. Fan, X.; White, I.M. Optofluidic microsystems for chemical and biological analysis. *Nat. Photonics* **2011**, *5*, 591–597. [CrossRef] [PubMed]

28. Cubillas, A.M.; Unterkofler, S.; Euser, T.G.; Etzold, B.J.; Jones, A.C.; Sadler, P.J.; Wasserscheid, P.; Russell, P.S.J. Photonic crystal fibres for chemical sensing and photochemistry. *Chem. Soc. Rev.* **2013**, *42*, 8629–8648. [CrossRef] [PubMed]

29. He, Z.; Tian, F.; Zhu, Y.; Lavlinskaia, N.; Du, H. Long-period gratings in photonic crystal fiber as an optofluidic label-free biosensor. *Biosens. Bioelectron.* **2011**, *26*, 4774–4778. [CrossRef] [PubMed]

30. Wu, C.; Tse, M.L.; Liu, Z.; Guan, B.O.; Zhang, A.P.; Lu, C.; Tam, H.Y. In-line microfluidic integration of photonic crystal fibres as a highly sensitive refractometer. *Analyst* **2014**, *139*, 5422–5429. [CrossRef] [PubMed]

31. Zhang, N.; Humbert, G.; Wu, Z.; Li, K.; Shum, P.P.; Zhang, N.M.; Cui, Y.; Auguste, J.L.; Dinh, X.Q.; Wei, L. In-line optofluidic refractive index sensing in a side-channel photonic crystal fiber. *Opt. Express* **2016**, *24*, 27674–27682. [CrossRef] [PubMed]

32. Kozma, P.; Kehl, F.; Ehrentreich-Forster, E.; Stamm, C.; Bier, F.F. Integrated planar optical waveguide interferometer biosensors: A comparative review. *Biosens. Bioelectron.* **2014**, *58*, 287–307. [CrossRef] [PubMed]

33. Yin, D.; Schmidt, H.; Barber, J.P.; Lunt, E.J.; Hawkins, A.R. Optical characterization of arch-shaped arrow waveguides with liquid cores. *Opt. Express* **2005**, *13*, 10564–10570. [CrossRef] [PubMed]

34. Testa, G.; Persichetti, G.; Sarro, P.M.; Bernini, R. A hybrid silicon-pdms optofluidic platform for sensing applications. *Biomed. Opt. Express* **2014**, *5*, 417–426. [CrossRef] [PubMed]

35. Testa, G.; Persichetti, G.; Bernini, R. Liquid core arrow waveguides: A promising photonic structure for integrated optofluidic microsensors. *Micromachines* **2016**, *7*, 47. [CrossRef]

36. Testa, G.; Huang, Y.; Sarro, P.M.; Zeni, L.; Bernini, R. High-visibility optofluidic mach-zehnder interferometer. *Opt. Lett.* **2010**, *35*, 1584–1586. [CrossRef] [PubMed]

37. Testa, G.; Huang, Y.; Sarro, P.M.; Zeni, L.; Bernini, R. Integrated silicon optofluidic ring resonator. *Appl. Phys. Lett.* **2010**, *97*, 131110. [CrossRef]

38. Dolatabady, A.; Asgari, S.; Granpayeh, N. Tunable mid-infrared nanoscale graphene-based refractive index sensor. *IEEE Sens. J.* **2017**, *18*, 569–574. [CrossRef]

39. Tan, Y.; He, R.; Cheng, C.; Wang, D.; Chen, Y.; Chen, F. Polarization-dependent optical absorption of MoS_2 for refractive index sensing. *Sci. Rep.* **2014**, *4*, 7523. [CrossRef] [PubMed]

40. Vollmer, F.; Braun, D.; Libchaber, A.; Khoshsima, M.; Teraoka, I.; Arnold, S. Protein detection by optical shift of a resonant microcavity. *Appl. Phys. Lett.* **2002**, *80*, 4057–4059. [CrossRef]

41. Zhi, Y.; Yu, X.C.; Gong, Q.; Yang, L.; Xiao, Y.F. Single nanoparticle detection using optical microcavities. *Adv. Mater.* **2017**, *29*. [CrossRef] [PubMed]

42. Luo, Y.; Chen, X.; Xu, M.; Chen, Z.; Fan, X. Optofluidic glucose detection by capillary-based ring resonators. *Opt. Laser Technol.* **2014**, *56*, 12–14. [CrossRef]

43. Scholten, K.; Fan, X.; Zellers, E.T. A microfabricated optofluidic ring resonator for sensitive, high-speed detection of volatile organic compounds. *Lab Chip* **2014**, *14*, 3873–3880. [CrossRef] [PubMed]

44. Zhang, S.X.; Wang, L.; Li, Z.Y.; Li, Y.; Gong, Q.; Xiao, Y.F. Free-space coupling efficiency in a high-q deformed optical microcavity. *Opt. Lett.* **2016**, *41*, 4437–4440. [CrossRef] [PubMed]

45. Lee, B.; Roh, S.; Park, J. Current status of micro-and nano-structured optical fiber sensors. *Opt. Fiber Technol.* **2009**, *15*, 209–221. [CrossRef]

46. Homola, J. Present and future of surface plasmon resonance biosensors. *Anal. Bioanal. Chem.* **2003**, *377*, 528–539. [CrossRef] [PubMed]

47. Barik, A.; Otto, L.M.; Yoo, D.; Jose, J.; Johnson, T.W.; Oh, S.H. Dielectrophoresis-enhanced plasmonic sensing with gold nanohole arrays. *Nano Lett.* **2014**, *14*, 2006–2012. [CrossRef] [PubMed]

48. Kang, S.; Lehman, S.E.; Schulmerich, M.V.; Le, A.P.; Lee, T.W.; Gray, S.K.; Bhargava, R.; Nuzzo, R.G. Refractive index sensing and surface-enhanced raman spectroscopy using silver-gold layered bimetallic plasmonic crystals. *Beilstein J. Nanotechnol.* **2017**, *8*, 2492–2503. [CrossRef] [PubMed]

49. Zhang, D.; Lu, Y.; Jiang, J.; Zhang, Q.; Yao, Y.; Wang, P.; Chen, B.; Cheng, Q.; Liu, G.L.; Liu, Q. Nanoplasmonic biosensor: Coupling electrochemistry to localized surface plasmon resonance spectroscopy on nanocup arrays. *Biosens. Bioelectron.* **2015**, *67*, 237–242. [CrossRef] [PubMed]

50. White, I.M.; Fan, X. On the performance quantification of resonant refractive index sensors. *Opt. Express* **2008**, *16*, 1020–1028. [CrossRef] [PubMed]

51. Sherry, L.J.; Chang, S.-H.; Schatz, G.C.; Van Duyne, R.P.; Wiley, B.J.; Xia, Y. Localized surface plasmon resonance spectroscopy of single silver nanocubes. *Nano Lett.* **2005**, *5*, 2034–2038. [CrossRef] [PubMed]

52. Jian, A.; Zhang, X.; Zhu, W.; Yu, M. Optofluidic refractometer using resonant optical tunneling effect. *Biomicrofluidics* **2010**, *4*, 043008. [CrossRef] [PubMed]

53. Chin, L.; Liu, A.; Lim, C.; Lin, C.; Ayi, T.; Yap, P. An optofluidic volume refractometer using fabry–pérot resonator with tunable liquid microlenses. *Biomicrofluidics* **2010**, *4*, 024107. [CrossRef] [PubMed]

54. St-Gelais, R.; Masson, J.; Peter, Y.-A. All-silicon integrated fabry–pérot cavity for volume refractive index measurement in microfluidic systems. *Appl. Phys. Lett.* **2009**, *94*, 243905. [CrossRef]

55. Trichet, A.A.; Dolan, P.R.; James, D.; Hughes, G.M.; Vallance, C.; Smith, J.M. Nanoparticle trapping and characterization using open microcavities. *Nano Lett.* **2016**, *16*, 6172–6177. [CrossRef] [PubMed]

56. Mader, M.; Reichel, J.; Hänsch, T.W.; Hunger, D. A scanning cavity microscope. *Nat. Commun.* **2015**, *6*, 7249. [CrossRef] [PubMed]

57. Chin, L.K.; Liu, A.Q.; Lim, C.S.; Zhang, X.M.; Ng, J.H.; Hao, J.Z.; Takahashi, S. Differential single living cell refractometry using grating resonant cavity with optical trap. *Appl. Phys. Lett.* **2007**, *91*, 243901. [CrossRef]

58. Chin, L.K.; Liu, A.Q.; Lim, C.S.; Lin, C.L.; Ayi, T.C.; Yap, P.H. An optofluidic volume refractometer using fabry-perot resonator with tunable liquid microlenses. *Biomicrofluidics* **2010**, *4*. [CrossRef] [PubMed]
59. Wang, Y.; Dostalek, J.; Knoll, W. Magnetic nanoparticle-enhanced biosensor based on grating-coupled surface plasmon resonance. *Anal. Chem.* **2011**, *83*, 6202–6207. [CrossRef] [PubMed]
60. Li, M.; Wu, X.; Liu, L.; Fan, X.; Xu, L. Self-referencing optofluidic ring resonator sensor for highly sensitive biomolecular detection. *Anal. Chem.* **2013**, *85*, 9328–9332. [CrossRef] [PubMed]
61. Xiao, Y.-F.; He, L.; Zhu, J.; Yang, L. Electromagnetically induced transparency-like effect in a single polydimethylsiloxane-coated silica microtoroid. *Appl. Phys. Lett.* **2009**, *94*, 231115. [CrossRef]
62. Lu, H.; Liu, X.; Mao, D.; Wang, G. Plasmonic nanosensor based on Fano resonance in waveguide-coupled resonators. *Opt. Lett.* **2012**, *37*, 3780–3782. [CrossRef] [PubMed]

micromachines

MDPI

Review

Optofluidic Technology for Water Quality Monitoring

Ning Wang [1,*], Ting Dai [1] and Lei Lei [2,*]

[1]   National Engineering Laboratory for Fiber Optic Sensing Technology, Wuhan University of Technology, Wuhan 430070, China; tingdai@whut.edu.cn
[2]   Shenzhen MacRitchie Technology Co., Ltd., Shen Zhen 518101, China
*   Correspondence: ningwang23@whut.edu.cn (N.W.); nm3rnd@gmail.com (L.L.)

Received: 25 February 2018; Accepted: 26 March 2018; Published: 1 April 2018

Abstract: Water quality-related incidents are attracting attention globally as they cause serious diseases and even threaten human lives. The current detection and monitoring methods are inadequate because of their long operation time, high cost, and complex process. In this context, there is an increasing demand for low-cost, multiparameter, real-time, and continuous-monitoring methods at a higher temporal and spatial resolution. Optofluidic water quality sensors have great potential to satisfy this requirement due to their distinctive features including high throughput, small footprint, and low power consumption. This paper reviews the current development of these sensors for heavy metal, organic, and microbial pollution monitoring, which will breed new research ideas and broaden their applications.

Keywords: optofluidic device; water quality; detection

1. Introduction

Water pollution has become one of the most pressing environmental problems in the world today [1–4]. Water research, especially water pollution analysis, has drawn attention from the general public. The analytical methods must be sensitive, accurate, high-speed, and automatic. Most detection technologies in modern analytical chemistry have been used in water pollution analysis, such as plasma emission spectrometry [5], atomic fluorescence spectrometry [6], gas chromatography–mass spectrometry [7], and high-performance liquid chromatography [8]. While researchers are still continuing to develop large-scale, sophisticated monitoring systems, there is an emerging trend towards portable, automated, continuous, simple, and fast detection devices [9–13].

Traditional methods to monitor water pollutants normally need to collect samples from a reservoir or watershed and then send them to laboratories for testing and a final analysis report. Long operation time may lead to inaccurate results because the water samples may undergo chemical, biological, and physical reactions during this period. Although certain on-site or in-line instruments are available to monitor pH, DO (dissolved oxygen), ORP (oxidation reduction potential), and various cations and anions of water in the market, their bulk size and high energy consumption limit their application. In addition, certain lab-based detection methods are difficult to package into an on-site system, especially for microbial monitoring, because of the operational complexity and requirement of an ultra-clean operating environment [14–21].

Microfluidics, also known as lab-on-a-chip or biochip, refers to techniques to control, operate, and detect fluids at microscopic dimensions. It can realize the basic functions of chemical and biological laboratories on a chip, with the goals of miniaturizing, automating, and integrating multifunction from processing to testing samples through the intersection of chemistry, micromachining, computers, electronics, materials science, and biology [22,23]. Optofluidics, defined as a fusion of optics and microfluidics, has recently become a hot research topic due to its advantages such as integration, miniaturization, and high precision, and it shows high potential in applications of biomedical and environmental sensing.

Although a few attempts to commercialize the related technology for the real-world application have been reported, such as eventlab from optiqua and parasitometer from Water Optics Technology, optofluidic devices for water pollution monitoring are mostly used in laboratories. At present, there is a serious lack of systematic reviews on its applications in the field. In this review, recent progress on optofluidic devices is reported with a focus on the various methods that are used to detect chemical, microbiological, and other ecological pollutants. Moreover, we highlight several commercial products and discuss their potential applications to water research and environment science.

2. Optofluidics for Online Water Quality Monitoring

Optofluidics, a new interdisciplinary combination of microfludics and optics, integrates fluid and optical elements, making it feasible to create innovative sensors with enhanced selectivity, adjustability, and compactness.

Optofluidic devices integrate microfluidic and optical components onto one chip and tie them together with high interaction efficiency between light and fluids, which is utilized for sensing applications [24]. Furthermore, using a continuous microchannel network to drive microfluids in and out of the handling system makes the optofluidic devices very attractive for environmental and biological threat detection, especially for continuous online measurement of trace amounts [25]. During the last two decades, many researchers have demonstrated numerous fundamental optofluidic elements such as optofluidic waveguides, including liquid core Anti-Resonant Reflecting Optical Waveguides (ARROWs) [26–28], Photonics Crystal Waveguides [26], Slot Waveguides [26,29,30], Liquid-Liquid Waveguides [31], Jet Waveguides [32], etc. and optofluidic devices including on-chip spectroscopy, an interferometer, and a resonator. Considering their potential application for detecting diverse chemical and biological components, it is important to realize water quality monitoring using compact, low-cost, low energy consumption and in-line optofluidic devices.

Because of the large-scale use of pesticides, heavy metal pollutants, and airborne metallic pollutants, aquatic ecosystems are threatened on a global scale [33]. The current water quality testing methods can be placed in three categories: (a) artificial fixed-point sampling and then laboratory off-line analysis; (b) portable instruments for on-site manual sampling test; and (c) continuous sampling from the fixed test site and automatic online analysis in real time. The manual fixed-point sampling has low efficiency, high cost, and is difficult to maintain. A portable instrument for on-site manual sampling is usually a single instrument to measure parameters with a limited range; often it has low efficiency and on-site calibration problems. The last method has high operating costs and can only obtain water quality testing results at a fixed site.

Optofluidic devices are good candidates for online continuous water quality monitoring because they can be integrated into different application fields, have reasonably low manufacturing cost and small chemical/energy consumption, which will minimize the hardware investment and operational costs.

According to the WHO drinking water safety guidelines [34], chemical and microbial pollutants are the main risks to water quality safety. In the following paragraphs, we present recent progress in research on chemical and microbial pollutants detection using optofluidic devices.

3. Chemical Pollutants Detection

The sources of chemical constituents in drinking water can be divided into the following three kinds of categories: natural origin from rocks/clays, and the geological environment and climate. Moreover, waste from factories and residences, agricultural processes, water governance, and substances relating to pesticides were also included [34]. More than 100 different chemical components were proven to be harmful for people's health; they can be classified into inorganic anions, heavy metals ions, and organic pollutants.

3.1. Inorganic Anions

3.1.1. Optical Absorption Change by Chemical Reaction

Optofluidic systems for nutrient determination are mostly based on spectrophotometric detection, using microfabrication techniques to integrate various optoelectronic components. For example, an optofluidic chip is integrated with a data transmitting device for phosphate detection in water [33]. The system can be equipped in different positions to conduct a full range of on-line searching for phosphate pollution in the target area with a detection limit of at most 0.3 mg/L. A Fabry–Pérot resonator, which enhances absorption, is used to detect marine phosphates. The optofluidic system can detect phosphate in real time and is of great significance for the detection and prediction of harmful algal blooms in the marine environment [35].

Zhu et al. [35] designed a combination technology of optofluidics with microscale resonators for the detection of phosphate. As shown in Figure 1a, the system was composed of a Fabry–Pérot microcavity with two parallel fiber facets coated with an Au film (Figure 1b) and a microchannel with a width of 250 μm. The water-soluble components, which were controlled by the flow rates, consisted of a phosphate solution, ascorbic acid solution, and a mixture of 12% ammonium molybdate solution, 80% concentrated sulfuric acid, and 8% antimony potassium tartrate solution. Compared to the traditional methods, the optofluidic device could shorten the detection time from 10 min to 6 s and increase the detection limit to 100 μmol/L.

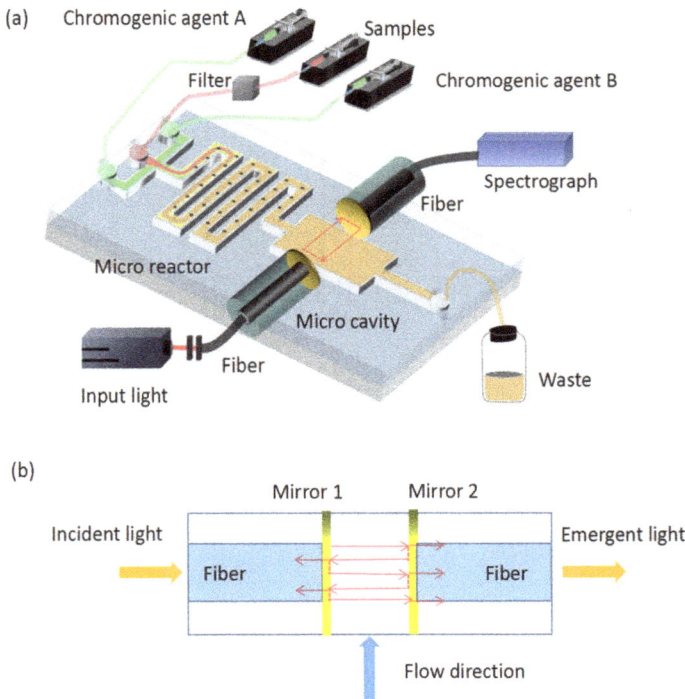

Figure 1. An optofluidic device based on the Fabry–Pérot resonator used to detect marine phosphate. (a) A 3D schematic of the setup. The microchannel area is 250 μm of width; the F–P cavity is 300 μm of length. (b) The microcavity is composed of two pairs of gold-coated optical fiber facets [35].

In addition, an optofluidic device for performing colorimetric measurement is presented in Figure 2a [36]. It allows a colorimetric reaction using a very small volume of sample and reagent (i.e., 20 μL), which are mixed in the T-mixer and connected by a serpentine channel where the interaction solution is provided to the emitter and photodetector for absorbance testing. Three inlets were designed on both sides of the T-mixer. The standard phosphate solutions and the water sample for calibration were introduced from one side. The reagent and two channels are for cleaning with deionized water (DI water) on the other side. In the microchannel, ammonium molybdate, $(NH_4)_6Mo_7O_{24}\cdot H_2O$, is reacted with ammonium metavanadate, NH_4VO_3, under acidic conditions. After being fully mixed and reacted, the vanadomolybdophosphoric acid complex, $(NH_4)_3PO_4NH_4VO_3\cdot16MoO_3$, was generated in the resulting solution, which presented a distinct yellow color, and therefore was accompanied by the strong absorbance below 400 nm. Also, the long-term reliability of the device was demonstrated in this work. From the experimental results, it has been proven to have a limit of detection (LOD) of 0.2 mg/L and a dynamic linear range of 0–50 mg/L. Due to the limited storage capacity, the optofluidic device could allow for >11,000 measurements.

Figure 2c,d show the fully assembled system [37]. With a Pelicase enclosure, five storage bottles and two waste bottles were fixed at the bottom. The black box between the waste bottles was the battery. Six solenoid pumps were fixed on the flipped top plate, as shown in Figure 2a. When we close the upper and lower plates, all the inlets and outlets can be accessed and connected with the microfluidic chip automatically. Moreover, the temperature sensor, LED, and photodiode were also configured on the top cover.

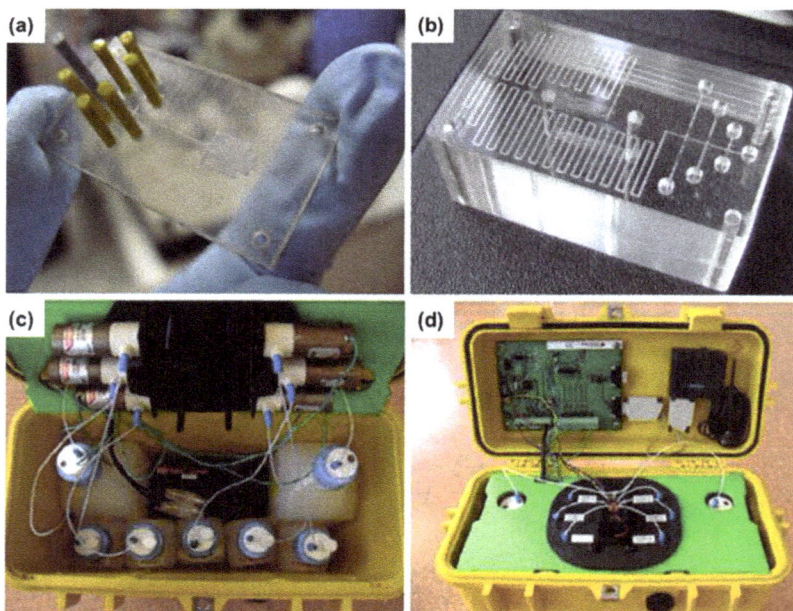

Figure 2. The optofluidic chips and assembled system used for phosphate detection. A high-sensitivity absorbance optical detection module is integrated with the optofluidic chip to monitor the colorimetric reaction of 20 μL volume reagent and sample. (**a**) The physical photo of microfluidic chip is described in this section. It shows a curved channel on the right side and six inlets on the left side with one outlet [36]. (**b**) Another chip redesigned to mix the sample and reagent, and the area used to detect is substituted with a cylindrical optical cuvette [36]; (**c**) an integrated device with bottles and battery [37]; (**d**) top board and cover showing electronic board and GSM modulator [37].

3.1.2. Fluorescence Quenching of Gold Nanoparticles in Water

For the detection of cyanide in tap water, drinking water, lake water, seawater, and industrial sewage, fluorescence quenching detection is developed using an optofluidic device. The integrated optofluidic device measures the fluorescence and colorimetric properties of gold nanoparticles mixed with the target water sample [38]. In these studies, the surface of Au NPs is modified by fluorescein isothIocyanate (fluorescent dye), which results in the fluorescence being quenched. Cyanide reacts with the Au NPs, forming a soluble gold/cyanide complex. The dye may then restore the fluorescence intensity. To stabilize Au NPs against high ionic strength, polysorbate 20 was used. When cyanide is present at the concentration larger than 150 µM, Au NPs' aggregation appears. In this way, the fluorescence could be detected at low concentrations of cyanide (LOQ of 10 µM). Thus, the colorimetric method is recommended for high concentration conditions (>150 µM).

The method described above has great sensitivity to cyanide compared to other common ions present in aqueous samples. The choice of detection method, fluorescence or colorimetric, depends on the amount of cyanide in the sample. In some cases with industrial water, the colorimetric method is more desirable.

3.2. Heavy Metal

3.2.1. Bioluminescence Inhibition of Specific Bacteria

The method of luminescent bacteria testing is mainly to suppress the photoexcitation of *Vibrio fischeri* cells in response to the toxic effects caused by heavy metal ions. *Vibrio fischeri* luminesces by generating luciferase, which catalyzes the oxidation of long-chain fatty aldehydes and reduces flavin mononucleotides, with free energy being released with luminescent of excited wavelength 490 nm. The traditional inhibition test is usually done by adding a volume of a luminescent bacterial suspension to the cuvette. The criterion for the test judgment was the attenuation of luminous intensity measured after 15 min and 30 min. In an optofluidic system, by processing in parallel the samples and standard liquid and measuring the luminescent intensity changes, the inhibition rate could be determined conveniently and rapidly. In Zhao's work [39], a novel optofluidic system based on the above principles and one-off toxicity assessments in ISO11384 was designed and characterized, as shown in Figure 3. The system worked via the following steps. First, the sample and luminescent bacterial suspension were continuously pumped into the micromixer at the same flow rate. After that, the mixture passed through a spiral microchannel to the viewing chamber. The optical measurement usually takes 20–30 min. Under fair control conditions of the Lab-on-a-Chip (LOC) system, the inhibition rate based on the luminescent intensity in the parallel microchamber could be calculated by observing and comparing the two steady states.

3.2.2. Color Change by Immunological Reaction

To achieve reliable observations of the bioluminescence inhibition, an observation chamber integrated with a highly sensitive photomultiplier tube (PMT) was designed in the optofluidic chip and its volume was optimized together with the droplet flow rate. Heavy metal ions (Hg^{2+}, Cd^{2+} and Pb^{2+}) in rivers, lakes, mineral and tap water could be detected and characterized by the antibody/antigen reaction in an optofluidic system. As reported by Zhou's group [40], a nitrocellulose membrane was covered separately by goat anti-mouse IgG antibody and the coating antigens Hg^{2+}-ITCBE-BSA, Cd^{2+}-ITCBE-BSA and Pb^{2+}-ITCBE-BSA(test-lines). Hg^{2+}, Cd^{2+}, and Pb^{2+} ions in water samples bind with the mAb/particle conjugates (specific mAb-labeled Au NPs), preventing their migration and capture by immobilized Hg^{2+}-ITCBE-BSA, Cd^{2+}-ITCBE-BSA, and Pb^{2+}-ITCBE-BSA. By using grayscale densitometry, the results show that the change of color density was inversely proportional to the concentration of metals.

This method allows the rapid (<10 min) and detection of multiple metal ions simultaneously with a high level of selectivity. The LODs for Hg^{2+}, Cd^{2+}, and Pb^{2+} were 8, 6, and 6 nM, respectively.

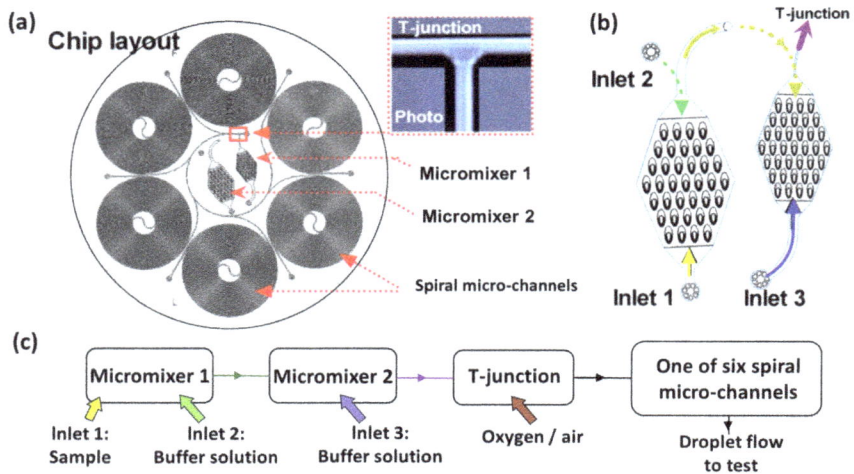

Figure 3. The bioluminescent cell-based optofluidic device for toxicant detection in water environment. *V. fischeri* is used as the toxicity indicator based on inhibition of luminescence; the device is designed to work continuously and shows high potential for home usage. (**a**) The optofluidic chip is composed of two counter-flow micromixers, a T-junction droplet generator, and six spiral microchannels; (**b**) two counter-flow micromixers and the flowing direction and mixing process; (**c**) the function of the different microstructures on the optofluidic chip [39].

3.2.3. Absorbance Change of Nanoparticles

Hg^{2+} ions can oxidize Ag nanoparticles (stabilized by sodium citrate) to Ag^+, causing the disintegration of nanoparticles. At the same time, Hg adsorbs on the Ag NPs, which results in the displacement of citrate molecules from the Ag NP surface and a decreasing negative charge of the particles. This leads to the enlargement of Ag NPs and a change in absorbance.

In an optofluidic device for Hg^{2+} detection [41], functionalized nanoparticles were fully mixed with the water sample in the microchannels and the absorbance of the solution was measured via a UV-Vis spectrophotometer. The measured range of Hg^{2+} concentration was 0.5–7 ppm, with precision RSDs of 3.24–4.53. The presence of Cu^{2+} ions enhances the sensitivity of Hg^{2+} detection: the LOD improves from 0.06 to 0.008 ppm.

3.3. Organic Pollutants

3.3.1. Micro-Ring Resonating Status Change by Immunological Reaction

Compared to inorganic pollutants in water, organic pollutants are more abundant. They affect ecosystems in a toxic and water-soluble manner and endanger human health. Therefore, the quantity of organic pollutants is an extremely important indicator of water pollution status. Due to their low content, these pollutants need to be pretreated in the early stage. The advantages of microfluidics include the ability to integrate the pretreatment and afterward detection, high extraction/enrichment efficiency, etc.

The optofluidic chip [42] shown in Figure 4 was used for sensitive monitoring of 2,4-Dichlorophenoxyacetic acid from different sources. Due to its excellent selection of antibodies, it has extremely high selectivity for 2,4-Dichlorophenoxyacetic acid. The optofluidic chip is fabrication by covalent immobilization of 2,4-D-bovine serum albumin (2,4-D-BSA) conjugate to an integrated microring resonator. It has a limit of detection (LOD) of 4.5 pg/mL and a quantitative range of 15–105 pg/mL. Its LOD was several orders of magnitude lower than conventional assays such as commercial enzyme linked immunosorbent assay (ELISA) kits, electrochemical impedance, fluorescence-labeled immunosensors, and APRs. More importantly, it requires a lower solution volume (~20 μL) than traditional ELISA methods (>100 μL).

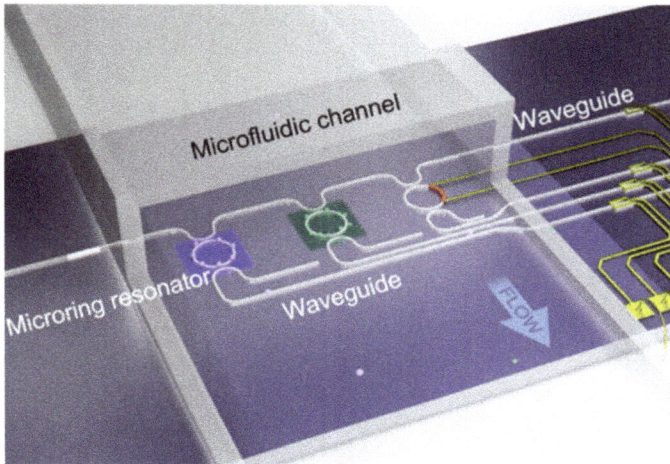

Figure 4. The optofluidic chip used to detect 2, 4-Dichlorophenoxyacetic [42].

3.3.2. Fluorescence Intensity Change by Immunological Reaction

Compared to traditional techniques such as ELISA, HPLC-UV, and invertebrate bioassay, an automated online optical biosensing system is rapid, sensitive, and high-frequency online-monitoring system for microcystin detection, and it is based on the principle of the total internal reflection fluorescence and flow indirect immunoassay. In Shi's research work [43], an innovative automated online optical biosensing system (AOBS) was designed to rapidly detect microcystin-LR (MC-LR), one of the most toxic cyanotoxins in water. MC-LR and anti-MC-LR-mAb within a certain concentration were premixed and pumped onto the chip surface. The laser was opened, at the same time exciting the surface-bound fluorescence and using an optical detector to record. The concentration of MC-LR, for which the LOD is 0.09 μg/L, ranges from 0.2 to 4 μg/L. The system has abilities of on-line detection and early warning response to water pollution, which has been successfully applied in Lake Tai in China. This novel system design was demonstrated with potential for detecting cyanobacteria and the reduction of pollution in fresh water, surface water, and drinking water.

Zhou et al. demonstrated that microfluidic systems also have important applications in the testing of chemical feedstocks [44]. In their study, an evanescent wave-excited immunological biosensor was designed to detect bisphenol A (BPA) with the detection limits of 0.03 mg/L. It is a portable device that only takes 20 min per assay cycle and can perform more than 300 assay cycles. This technique has low LOD, high sensitivity, and excellent selectivity in a real water environmental matrix. Due to the reconfiguration of the planar optical waveguide microchip, it is possible to have high testing frequency for water pollution detection.

4. Microbial Pollutants Detection

Microbial pollutants such as pathogenic bacteria, viruses, and parasites in water samples are very common [34]. Their typical size ranges from tens of nanometer (virus) to tens of micrometer (protozoa), which is far smaller than the eye can see. The current lab-based test methods are normally time-consuming (24 h to 7 d) and involve complicated procedures, normally including water sample filtration, microbial purification, culture, and observation by microscopic methods. Flow cytometry is an alternative method for their determination. However, its equipment is expensive, bulky, and requires special operation, and therefore it is not suitable for on-site and continuous monitoring.

Based on their size, the microorganisms in water bodies belong to the scope of particulate organic carbon. Their species groups can reflect the ecological characteristics of water bodies and the pollution status, which is the routine monitoring index in water ecological surveys. The emergence of an optofluidic device based on sheath flow fluid control makes the integration, miniaturization, automation, and portability of the instrument possible. In addition, the combination of nucleic acid or immunological bioassay methods and immunomagnetic separation technology for the detection of waterborne pathogens is of great significance.

4.1. Microdroplet Scattering Change by Bio-Reaction

Traditionally, the marking method was used for bacteriophage detection, which has the disadvantages of complex, time-consuming steps and a low recovery rate. In Yu's research work [45], a label-free method was employed to detect phages, here named droplet optofluidic imaging. Phage infection reaction assays used droplets containing host cells as a carrier due to the higher proportion of phages and host cells in the droplets. Optofluidic imaging relied on the variety of the effective refractive index of the growth rate of infected host cells in the droplet, presents high sensitivity, and can even detect a single *E. coli* cell.

Figure 5 shows the droplet optofluidic imaging system for phage monitoring, in which microdroplets were utilized as the reaction containers and light scattering was captured for optical imaging. When light is induced onto the droplets, including both the host cells and the phage, the light forms a scattering pattern on the image plane. The host cell growth rate was effectively correlated and the information can be taken from the photo-fluidic imaging signal. The optofluidic imaging system not only has high potential for monitoring water quality, especially for drinking water, but can also be used in clinical diagnosis and pathogenic bacteria detection in the food industry.

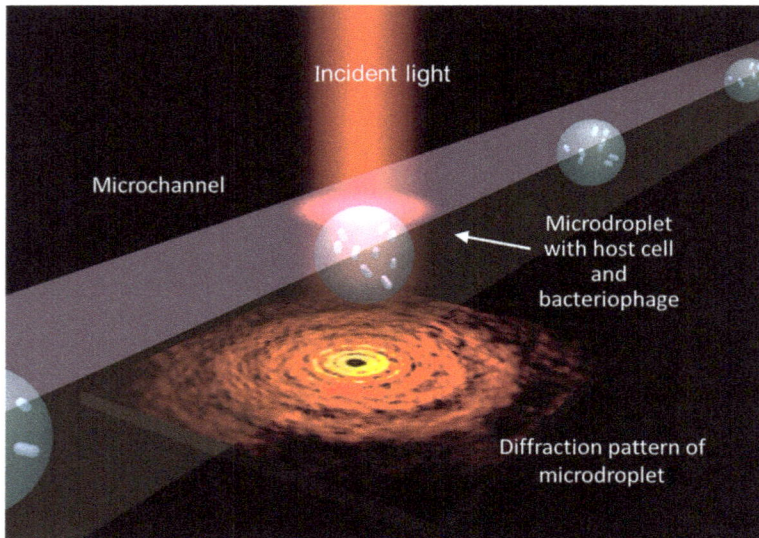

Figure 5. Schematic diagram of the optofluidic imaging based on microdroplets, co-cultured bacteriophage, and host cells (*E. coli*) [45].

4.2. Bacteria Enrichment and PCR Detection

The enrichment and detection of *E. coli* O157:H7 in aqueous samples is also of significance. In Dharmasiri's research [46], a microfluidic chip was fabricated to isolate and enrich *E. coli* O157:H7 cells with low abundance (<100 cells/mL) from water samples. The microchip could process eight kinds of samples independently or a sample pumped into eight different parallel microchannels to increase productive forces. After enrichment, cells were released and enumerated by utilizing benchtop real-time quantitative Polymerase Chain Reaction (PCR). The recovery of target cells from water samples was detected to be about ~72%, and the limit-of-detection reached six colony forming units (CFU) by utilizing slt1 gene as a reporter.

Lake and waste water samples can be analyzed with this device. Both the fabrication and operation processes were simple, which makes this kind of device attractive and competitive for the detection of pathogenic species from water samples, including bacterial pathogens with quite low frequencies.

4.3. Virus PCR Dection

An optofluidic quantitative PCR (MFQPCR) system was designed to detect 11 major human viral pathogens simultaneously [47]. The researchers collected samples from the Sapporo Sewage Treatment Plant in Japan to validate the presence of various viruses in the MFQPCR system. By using this system, the viral quantities obtained were nearly the same as in the traditional method. Thus, the direct and quantitative information of viral pathogens can be obtained from the MFQPCR system, which is used for risk assessments.

4.4. Automatic Microscopic Identification of Parasites

Optical microscopy is widely utilized as an analytical tool for biomedical and biological applications. In 2006, Heng et al. reported that the novel technique of optofluidic microscopy (OFM) [48] combines microfluidics and optics for low-cost microscopy imaging with high resolution. Instead of using lots of lenses and other optics of conventional optical microscopes, the OFM performs an imaging translation/scanning of the target biological specimen through a gold-plated array of holes at the bottom of the microchannels.

Lee et al. designed and set up a complete on-chip OFM system that was designed to image *Caenorhabditis elegans*, as shown in Figure 6 [49]. The OFM system can achieve high-resolution microscopic images beyond the sensor pixel size limitation by reconstruction of raw acquired images using a special image processing algorithm. OFM can also be used for imaging *Giardia lamblia* trophozoites and cysts, a parasite species causing infectious diseases and commonly found in poor-quality water. OFM has achieved a focal plane resolution of 800 nm. The sub-cellular morphology information is able to identify specific waterborne protozoan parasites. This study shows that the OFM technology can potentially be applied to develop autonomous, low-cost, and highly compact water analysis systems to monitor water quality in hazardous environments and underdeveloped countries.

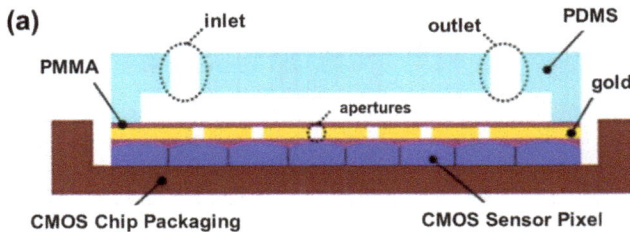

Figure 6. *Cont.*

Figure 6. On chip OFM system for the detection of protozoan parasites in drinking water. Detection of disease-causing parasite species like *Caenorhabditis elegans* and *Giardia lamblia* was demonstrated using OFM technology. (**a**) Cross section view of the OFM device; (**b**) photo of the fabricated on-chip OFM device [49].

5. Discussion and Outlook

As optofluidic devices for water quality monitoring develop rapidly and show high potential for applications, several national agencies have funded projects for the technology's development and commercialization [50–54]. Many prototypes and quasi-ready products have been developed and validated in real-world applications. Autonomous lab-on-a-chip biosensor-based equipment has been prototyped and its online chemical detection performance validated on the coast of Sagrada of Greece [52]. A label-free detection system for rapid protozoa detection in drinking water has been prototyped and its online *Cryptosporidium* detection performance validated in Singapore [51]. Miyake's group at Tokyo University demonstrated 3D-printing-based optofluidic analyzers for monitoring water quality, which have been adopted by Toshiba. More and more mature optofluidic devices have been applied in the water industry.

Some companies have also successfully commercialized their optofluidic devices for online drinking water quality monitoring. Optiqua has commercialized Eventlab as a real-time water chemical event sensor to monitor drinking water network safety using an optofluidic based on the Mach–Zehnder interferometer, as shown in Figure 7a. Their products have been successfully applied by several utility companies worldwide [55,56]. Figure 7b shows a photograph of the optofluidic system installed in water plants for monitoring sodium chloride and so on.

Refractive index (RI) is used as an indicator to monitor the total chemical concentration and variance. A sudden RI change in the drinking water network will set off alarms for utility companies, prompting them to act immediately to prevent a second cross-contamination in the network. Their products have been successfully applied by several utility companies worldwide [57].

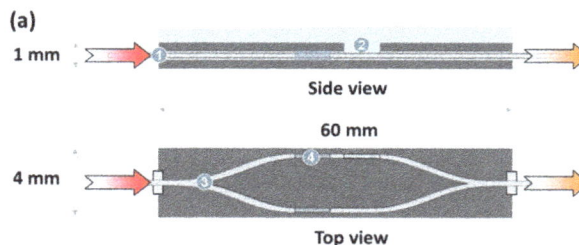

Figure 7. *Cont.*

Figure 7. EventLab for online chemical event sensing. A new monitoring concept to continuously monitor the refractive index (RI) in real time is developed. RI is considered to be an indicator of the full spectrum of possible chemical contaminants and EventLab can be used as a highly sensitive (PPM level) generic optical sensor and an early-stage chemical contaminant warning system. (**a**) Schematic of the MZI chip used for measuring RI; (**b**) photograph of the optofluidic system installed in water plants for monitoring sodium chloride and so on [57].

6. Conclusions

The application of optofluidics in water quality testing has become a mainstream trend in understanding water quality. Though impressive results and obvious significance have been reported, the promise of optofluidic systems for water industry will be realized only through continuous use of new technologies and converting these fundamental principles into prospective real-world applications. With the development of different types of optofluidic systems, real-time detection can be universal, and will make remarkable contributions to the water detection and environment research.

Acknowledgments: This work is supported by the National Science Foundation of China (No. 61605148) and the Fundamental Research Funds for the Central University (No. 2017IVB019 and No. 2017IVA081).

Author Contributions: Ning Wang planned the configuration and key topics and wrote the manuscript. Ning Wang and Lei Lei revised the paper for language and quality. Ting Dai surveyed the literature, prepared the figures and obtained permissions, and drafted the manuscript.

Conflicts of Interest: The authors declare no conflict of interest.

References

1. Duić, N.; Guzović, Z.; Kafarov, V.; Klemeš, J.J.; vad Mathiessen, B.; Yan, J. Sustainable development of energy, water and environment systems. *Appl. Energy* **2013**, *101*, 3–5. [CrossRef]
2. Baker, L.A. *The Water Environment of Cities*; Springer: New York, NY, USA, 2009; ISBN 9780387848907.
3. Wang, Q.; Yang, Z. Industrial water pollution, water environment treatment, and health risks in China. *Environ. Pollut.* **2016**, *218*, 358–365. [CrossRef] [PubMed]
4. Alherifiere, D. Water and environment. *Water Int.* **1980**, *5*, 4–8. [CrossRef]
5. Smith, C.L.; Motooka, J.M.; Willson, W.R. Analysis of trace metals in water by inductively coupled plasma emission spectrometry using sodium dibenzyldithiocarbamate for preconcentration. *Anal. Lett.* **1984**, *17*, 1715–1730. [CrossRef]
6. Sayago, A.; Bcltrân, R.; Gômez-Ariza, J.L. Hydride generation atomic fluorescence spectrometry (HG-AFS) as a sensitive detector for Sb(in) and Sb(v) speciation in water. *J. Anal. At. Spectrom.* **2000**, *15*, 423–428. [CrossRef]

7. Safarova, V.I.; Sapelnikova, S.V.; Djazhenko, E.V.; Teplova, G.I.; Shajdulina, G.F.; Kudasheva, F.K. Gas chromatography-mass spectrometry with headspace for the analysis of volatile organic compounds in waste water. *J. Chromatogr. B Anal. Technol. Biomed. Life Sci.* **2004**, *800*, 325–330. [CrossRef]

8. Pinto, G.M.; Jardim, I.C. Use of solid-phase extraction and high-performance liquid chromatography for the determination of triazine residues in water: Validation of the method. *J. Chromatogr. A* **2000**, *869*, 463–469. [CrossRef]

9. Toivanen, T.; Koponen, S.; Kotovirta, V.; Molinier, M.; Chengyuan, P. Water quality analysis using an inexpensive device and a mobile phone. *Environ. Syst. Res.* **2013**, *2*, 9. [CrossRef]

10. Şener, Ş.; Şener, E.; Davraz, A. Evaluation of water quality using water quality index (WQI) method and GIS in Aksu River (SW-Turkey). *Sci. Total Environ.* **2017**, *584–585*, 131–144. [CrossRef] [PubMed]

11. Chang, K.; Gao, J.L.; Wu, W.Y.; Yuan, Y.X. Water quality comprehensive evaluation method for large water distribution network based on clustering analysis. *J. Hydroinformatics* **2011**, *13*, 390–400. [CrossRef]

12. Jiang, Y.; Nan, Z.; Yang, S. Risk assessment of water quality using Monte Carlo simulation and artificial neural network method. *J. Environ. Manag.* **2013**, *122*, 130–136. [CrossRef] [PubMed]

13. Deng, W.; Wang, G. A novel water quality data analysis framework based on time-series data mining. *J. Environ. Manag.* **2017**, *196*, 365–375. [CrossRef] [PubMed]

14. Richardson, S.D. Water analysis: Emerging contaminants and current issues. *Anal. Chem.* **2009**, *81*, 4645–4677. [CrossRef] [PubMed]

15. Fawell, J.; Nieuwenhuijsen, M.J. Contaminants in drinking water. *Br. Med. Bull.* **2003**, *68*, 199–208. [CrossRef] [PubMed]

16. Richardson, S.D.; Kimura, S.Y. Water analysis: Emerging contaminants and current issues. *Anal. Chem.* **2016**, *88*, 546–582. [CrossRef] [PubMed]

17. Petrović, M.; Gonzalez, S.; Barceló, D. Analysis and removal of emerging contaminants in wastewater and drinking water. *TrAC Trends Anal. Chem.* **2003**, *22*, 685–696. [CrossRef]

18. Richardson, S.D.; Ternes, T.A. Water analysis: Emerging contaminants and current issues. *Anal. Chem.* **2011**, *83*, 4616–4648. [CrossRef] [PubMed]

19. Richardson, S.D.; Ternes, T.A. Water analysis: Emerging contaminants and current issues. *Anal. Chem.* **2014**, *86*, 2813–2848. [CrossRef] [PubMed]

20. Mao, S.; Chang, J.; Zhou, G.; Chen, J. Nanomaterial-enabled rapid detection of water contaminants. *Small* **2015**, *11*, 5336–5359. [CrossRef] [PubMed]

21. Schriks, M.; Heringa, M.B.; van der Kooi, M.M.E.; de Voogt, P.; van Wezel, A.P. Toxicological relevance of emerging contaminants for drinking water quality. *Water Res.* **2010**, *44*, 461–476. [CrossRef] [PubMed]

22. Nahavandi, S.; Baratchi, S.; Soffe, R.; Tang, S.-Y.; Nahavandi, S.; Mitchell, A.; Khoshmanesh, K. Microfluidic platforms for biomarker analysis. *Lab Chip* **2014**, *14*, 1496–1514. [CrossRef] [PubMed]

23. Thorsen, T.; Maerkl, S.J.; Quake, S.R. Microfluidic large-scale integration. *Science* **2002**, *298*, 580–584. [CrossRef] [PubMed]

24. Testa, G.; Persichetti, G.; Bernini, R. Optofluidic approaches for enhanced microsensor performances. *Sensors* **2015**, *15*, 465–484. [CrossRef] [PubMed]

25. Persichetti, G.; Testa, G.; Bernini, R. High sensitivity UV fluorescence spectroscopy based on an optofluidic jet waveguide. *Opt. Express* **2013**, *21*, 24219. [CrossRef] [PubMed]

26. Fan, X.; White, I.M.; Shopova, S.I.; Zhu, H.; Suter, J.D.; Sun, Y. Sensitive optical biosensors for unlabeled targets: A review. *Anal. Chim. Acta* **2008**, *620*, 8–26. [CrossRef] [PubMed]

27. Testa, G.; Persichetti, G.; Sarro, P.M.; Bernini, R. A hybrid silicon-PDMS optofluidic platform for sensing applications. *Biomed. Opt. Express* **2014**, *5*, 417. [CrossRef] [PubMed]

28. Yin, D.; Deamer, D.W.; Schmidt, H.; Barber, J.P.; Hawkins, A.R. Integrated optical waveguides with liquid cores. *Appl. Phys. Lett.* **2004**, *85*, 3477–3479. [CrossRef]

29. Almeida, V.R.; Xu, Q.; Barrios, C.A.; Lipson, M. Guiding and confining light in void nanostructure. *Opt. Lett.* **2004**, *29*, 1209. [CrossRef] [PubMed]

30. Sun, Y.; Fan, X. Optical ring resonators for biochemical and chemical sensing. *Anal. Bioanal. Chem.* **2011**, *399*, 205–211. [CrossRef] [PubMed]

31. Wolfe, D.B.; Conroy, R.S.; Garstecki, P.; Mayers, B.T.; Fischbach, M.; Paul, K.E.; Prentiss, M.; Whitesides, G.M. Dynamic control of liquid-core/liquid-cladding optical waveguides. *Proc. Natl. Acad. Sci. USA* **2004**, *101*, 12434–12438. [CrossRef] [PubMed]

32. Persichetti, G.; Testa, G.; Bernini, R. Optofluidic jet waveguide enhanced Raman spectroscopy. *Sens. Actuators B Chem.* **2015**, *207*, 732–739. [CrossRef]

33. Bartram, J.; Ballance, R. *Water Quality Monitoring—A Practical Guide to the Design and Implementation of Freshwater Quality Studies and Monitoring Programmes*; CRC Press: Boca Raton, FL, USA, 1996.

34. Gorchev, H.G.; Ozolins, G. WHO guidelines for drinking-water quality. *WHO Chron.* **2011**, *38*, 104–108. [CrossRef]

35. Zhu, J.M.; Shi, Y.; Zhu, X.Q.; Yang, Y.; Jiang, F.H.; Sun, C.J.; Zhao, W.H.; Han, X.T. Optofluidic marine phosphate detection with enhanced absorption using a Fabry–Pérot resonator. *Lab Chip* **2017**, *17*, 4025–4030. [CrossRef] [PubMed]

36. Cleary, J.; Maher, D.; Diamond, D. Development and deployment of a microfluidic platform for water quality monitoring. In *Smart Sensors Real-Time Water Quality Monitoring*; Springer: Berlin/Heidelberg, Germany, 2013; Volume 4, pp. 125–148.

37. Slater, C.; Cleary, J.; McGraw, C.M.; Yerazunis, W.S.; Lau, K.T.; Diamond, D. Autonomous field-deployable device for the measurement of phosphate in natural water. *Proc. SPIE* **2007**, *6755*, 67550L. [CrossRef]

38. Cheng, C.; Chen, H.Y.; Wu, C.S.; Meena, J.S.; Simon, T.; Ko, F.H. A highly sensitive and selective cyanide detection using a gold nanoparticle-based dual fluorescence-colorimetric sensor with a wide concentration range. *Sens. Actuators B Chem.* **2016**, *227*, 283–290. [CrossRef]

39. Zhao, X.; Dong, T. A microfluidic device for continuous sensing of systemic acute toxicants in drinking water. *Int. J. Environ. Res. Public Health* **2013**, *10*, 6748–6763. [CrossRef] [PubMed]

40. Zhou, Y.; Li, Y.S.; Meng, X.Y.; Zhang, Y.Y.; Yang, L.; Zhang, J.H.; Wang, X.R.; Lu, S.Y.; Ren, H.L.; Liu, Z.S. Development of an immunochromatographic strip and its application in the simultaneous determination of Hg(II), Cd(II) and Pb(II). *Sens. Actuators B Chem.* **2013**, *183*, 303–309. [CrossRef]

41. Jarujamrus, P.; Amatatongchai, M.; Thima, A.; Khongrangdee, T.; Mongkontong, C. Selective colorimetric sensors based on the monitoring of an unmodified silver nanoparticles (AgNPs) reduction for a simple and rapid determination of mercury. *Spectrochim. Acta Part A Mol. Biomol. Spectrosc.* **2015**, *142*, 86–93. [CrossRef] [PubMed]

42. Feng, X.; Zhang, G.; Chin, L.K.; Liu, A.Q.; Liedberg, B. Highly sensitive, label-free detection of 2,4-dichlorophenoxyacetic acid using an optofluidic chip. *ACS Sens.* **2017**, *2*, 955–960. [CrossRef] [PubMed]

43. Shi, H.C.; Song, B.D.; Long, F.; Zhou, X.H.; He, M.; Lv, Q.; Yang, H.Y. Automated online optical biosensing system for continuous real-time determination of microcystin-LR with high sensitivity and specificity: Early warning for cyanotoxin risk in drinking water sources. *Environ. Sci. Technol.* **2013**, *47*, 4434–4441. [CrossRef] [PubMed]

44. Zhou, X.-H.; Liu, L.-H.; Xu, W.-Q.; Song, B.-D.; Sheng, J.-W.; He, M.; Shi, H.-C. A reusable evanescent wave immunosensor for highly sensitive detection of bisphenol A in water samples. *Sci. Rep.* **2014**, *4*, 17–20. [CrossRef]

45. Yu, J.Q.; Huang, W.; Chin, L.K.; Lei, L.; Lin, Z.P.; Ser, W.; Chen, H.; Ayi, T.C.; Yap, P.H.; Chen, C.H.; et al. Droplet optofluidic imaging for λ-bacteriophage detection via co-culture with host cell *Escherichia coli*. *Lab Chip* **2014**, *14*, 3519–3524. [CrossRef] [PubMed]

46. Dharmasiri, U.; Witek, M.A.; Adams, A.A.; Osiri, J.K.; Hupert, M.L.; Bianchi, T.S.; Roelke, D.L.; Soper, S.A. Enrichment and detection of *Escherichia coli* O157:H7 from water samples using an antibody modified microfluidic chip. *Anal. Chem.* **2010**, *82*, 2844–2849. [CrossRef] [PubMed]

47. Ishii, S.; Kitamura, G.; Segawa, T.; Kobayashi, A.; Miura, T.; Sano, D.; Okabe, S. Microfluidic quantitative PCR for simultaneous quantification of multiple viruses in environmental water samples. *Appl. Environ. Microbiol.* **2014**, *80*, 7505–7511. [CrossRef] [PubMed]

48. Heng, X.; Erickson, D.; Baugh, L.R.; Yaqoob, Z.; Sternberg, P.W.; Psaltis, D.; Yang, C. Optofluidic microscopy—A method for implementing a high resolution optical microscope on a chip. *Lab Chip* **2006**, *6*, 1274–1276. [CrossRef] [PubMed]

49. Lee, L.M.; Cui, X.; Yang, C. The application of on-chip optofluidic microscopy for imaging *Giardia lamblia* trophozoites and cysts. *Biomed. Microdevices* **2009**, *11*, 951–958. [CrossRef] [PubMed]

50. Microfluidic Sensors for In-Line Water Monitoring Applications. Available online: https://www.sbir.gov/sbirsearch/detail/10875 (accessed on 29 March 2018).

51. Microfabricated, Low-Cost, High-Sensitivity Chlorine and pH Sensor Systems for Water Quality Monitoring. Available online: https://ic-impacts.com/portfolio-posts/microfabricated-low-cost-high-sensitivity-chlorine-and-ph-sensor-systems-for-water-quality-monitoring/ (accessed on 29 March 2018).

52. Real Time Monitoring of SEA Contaminants by an Autonomous Lab-on-a-Chip Biosensor. Available online: https://cordis.europa.eu/project/rcn/111294_en.html (accessed on 29 March 2018).

53. Final Report: Development of Mobile Self-Powered Sensors for Potable Water Distribution. Available online: https://cfpub.epa.gov/ncer_abstracts/index.cfm/fuseaction/display.highlight/abstract/9458/report/F (accessed on 29 March 2018).

54. Platform Realising the Cloud-to-Things Continuum Concept. Available online: http://www.proteussensor.eu/ (accessed on 1 December 2017).

55. Nolan, B.T.; Hitt, K.J.; Ruddy, B.C. Probability of nitrate contamination of recently recharged groundwaters in the conterminous United States. *Environ. Sci. Technol.* **2002**, *36*, 2138–2145. [CrossRef] [PubMed]

56. Van den Broeke, J. *The Benefits of Using Refractive Index for Water Quality Monitoring in Distribution Networks*; Optiqua Technologies: Richmond, VIC, Australia, 2014.

57. Van Wijlen, M.A.B.; Koerkamp, M.K.; Xie, R.J.; Puah, A.N.; van Delft, W.; Bajema, B.; Verhoef, J.W. Innovative sensor technology for effective online water quality monitoring. In Proceedings of the 4th Singapore International Water Week, Singapore, 4–8 July 2011.

micromachines

MDPI

Review

High-Throughput Optofluidic Acquisition of Microdroplets in Microfluidic Systems

Zain Hayat and Abdel I. El Abed *

Laboratoire de Photonique Quantique et Moléculaire, UMR 8537, Ecole Normale Supérieure Paris Saclay, CentraleSupélec, CNRS, Université Paris-Saclay, 61 avenue du Président Wilson, 94235 Cachan, France; zain.hayat@ens-paris-saclay.fr
* Correspondence: abdel.el-abed@ens-paris-saclay.fr; Tel.: +33-147-405-562

Received: 27 February 2018; Accepted: 4 April 2018; Published: 14 April 2018

Abstract: Droplet optofluidics technology aims at manipulating the tiny volume of fluids confined in micro-droplets with light, while exploiting their interaction to create "digital" micro-systems with highly significant scientific and technological interests. Manipulating droplets with light is particularly attractive since the latter provides wavelength and intensity tunability, as well as high temporal and spatial resolution. In this review study, we focus mainly on recent methods developed in order to monitor real-time analysis of droplet size and size distribution, active merging of microdroplets using light, or to use microdroplets as optical probes.

Keywords: microfluidics; droplets; optofluidics

---

1. Introduction

The advent of segmented phase flow in microfluidic systems nearly two decades ago gave rise to the development of an important sub-field of microfluidics known as droplet microfluidics [1,2]. This technology allows for the fabrication and manipulation of millions of highly monodisperse microdroplets, each of which may be regarded as an independent micro-reactor [3–11]. The combination of the high flexibility of microfluidics and the compartmentalization of reagents in droplets at high throughput provides powerful automated tools for optimizing chemical synthesis and the development of rapid and low-cost digital assays. Droplets content can be incubated, split, merged, analyzed or sorted at kHz rates, which has proven, for instance, to be a powerful tool to find mutants of genes among a very large population of wild genes [12–18].

Optofluidics is another fast-growing research field dedicated to the study of the interaction of light with discrete volumes of liquids in microfluidic systems. One may remark that, besides the present themed collection dedicated to optofluidics, two previous themed collections have recently been published in *Lab on a Chip* (in 2013 and 2016) [19–32], as well as a series of international conferences held annually in China since 2011. The combination of droplet microfluidics and optofluidics, coined hereafter as droplet optofluidics, offers many prospects spanning many academic and industrial fields in biology, chemistry, physics, material, and interface sciences [33–51]. For instance, it enabled the identification of very rare gene sequences [12–18], screening of cells or bacteria [52], membrane proteins inhibitors screening [53], coupled optical lab-on-chip platform with small angle X-ray scattering (SAXS) [54], and engineering microparticles for photonics applications [55,56], on-chip multiphasic tunable grating [57], reconfigurable droplet grating [58], fluidic Michelson interferometer [59], droplet grating with polydimethylsiloxane (PDMS) air-lens waveguide setup [60], 3D and 4D optically fabricated complex geometries [61,62], reconfigurable compound micro-lenses [63], and countless other possibilities. Previous examples may be considered to be the state-of-the-art, and the domain is flourishing day-by-day to new trends.

Droplet size and droplet size distribution are among the most relevant and challenging characteristics of droplets, and can affect their use for highly quantitative analysis and biological assays. Various methods have been developed in order to monitor the size and size distribution of large populations of droplets in real-time. Some methods employ expensive equipment, such as dynamic light scattering [40], automated scanning electron microscopy [41], acoustic attenuation spectroscopy [42,43], and capillary hydrodynamics [44,45]. New methods based on the state-of-the-art microscopy have also been recently reported in the literature, such as image processing [46,64], real-time on-chip imaging and droplet-sorting systems based on real shape recognition methods [65,66], a coupled bright-field and fluorescence multi-imaging flow cytometer [67], and advanced digital acquisition [47]. In this review, we will focus on the developed optofluidics methods for monitoring and acquiring droplets size and size distribution. We will also tackle some of the engineering as well as crucial aspects of optofluidics methods for droplet manipulation.

2. Basics of Droplet Microfluidics Technology

Several materials have been used for the fabrication of microfluidic systems, each of which has its advantages and drawbacks. Silicon was first used for the development of microfluidic chips [68,69]. The reason for selecting silicon is evident from its inert behavior in regards to a wide range of chemical compounds. Being opaque (a major drawback), silicon was soon replaced by glass, which is not only chemically inert, but also transparent [70,71]. The development of glass microfluidic systems requires complicated design protocols. It is worth noting that many materials may be used for microfluidic systems, including thermoplastic [72,73], ceramic laminated sheets [74], thermoset polyester [75], polystyrene (PS) [76,77], poly-methyl-methacrylate (PMMA) [78,79] and polycarbonate (PC) [80,81]. However, polydimethylsiloxane (PDMS) is one of the most widely used polymers for the fabrication of reliable and cost-effective microfluidic devices [82,83]. Among the numerous advantages of PDMS to microfluidics, one may note for instance its excellent optical transparency, easy processing, low cost, mechanical flexibility, long-term stability, biocompatibility and low toxicity, chemical inertness, etc. Nevertheless, PDMS has one main drawback: its propensity for swelling in the presence of low molecular weight organic solvents such as acetone, ethanol, chloroform, etc., which may impede some applications requiring the use of low molecular weight organic solvents, for instance.

In order to generate highly monodisperse droplets, different designs have been developed. Some of them were developed at the very beginning of droplet microfluidics technology, such as T-junction [84–87], flow-focusing [32,37,39,47,65,88–91], co-flow [92–95], or glass capillary droplet generator [70,71,96,97]. New commercial platforms using micro-pipetting [98,99] are also available; for example, rotAXYS ® (Cetoni) and Dropix® (Dolomite). Nevertheless, T-junction and flow-focusing are the most commonly used droplet generators. They enable a continuous carrier oil flow to periodically slice, at high throughput, a second phase flow (dispersed phase) into tiny droplets at the nozzle region of the microfluidic device (see Figure 1d) [100]. The success of such an operation and the features of the fabricated droplets depend on parameters like the non-miscibility of the two fluids, flow rates (or flow velocity, u) of the fluids, interfacial tension (γ) of the two fluids, viscosity (η) of the carrier oil phase, the microchip nozzle size, etc. The mechanism lying behind the formation of highly monodisperse droplets in microfluidic devices is governed by a subtle balance between capillary forces minimizing the interfacial energy between the two fluids (through the formation of spherical droplets) and viscous forces acting on the dispersed phase during the deformation of the interface and the formation of the droplets. The result of a such balance may be evaluated using the dimensionless capillary number C_a, which is defined as the ratio between the viscous forces and the capillary forces, or $C_a \sim \dfrac{\eta u}{\gamma}$. Depending on the value of the capillary number, three different flow regimes are observed: dripping, jetting, and parallel flow. The first regime, where monodisperse periodic microdrops are produced near the microfluidic device nozzle, is obtained when capillary forces dominate viscous forces (i.e., when $C_a << 1$). The second regime is observed when viscous forces become comparable to capillary forces ($C_a \sim 1$). This regime is characterized by a long undulating jet, which breaks far downstream from the nozzle into polydisperse droplets. The third regime corresponds to the case where the two fluids flow continuously side by side, and is observed for high flow rates and/or highly viscous fluids (i.e., when $C_a \gg 1$) [101–104].

To serve successfully as independent microreactors, droplets should obey at least the following specifications: (i) high monodispersity; (ii) long-term stability; (iii) absence of cross-contamination between droplets; and (iv) for biomedical applications, their content and the surrounding medium should be biocompatible. All these specifications can be achieved by using a perfluorinated oil as a carrier fluid and a perfluorinated surfactant as a stabilizing droplet agent [105–108]. Indeed, perfluorinated compounds are chemically inert and mix neither with aqueous solutions nor with organic solvent solutions.

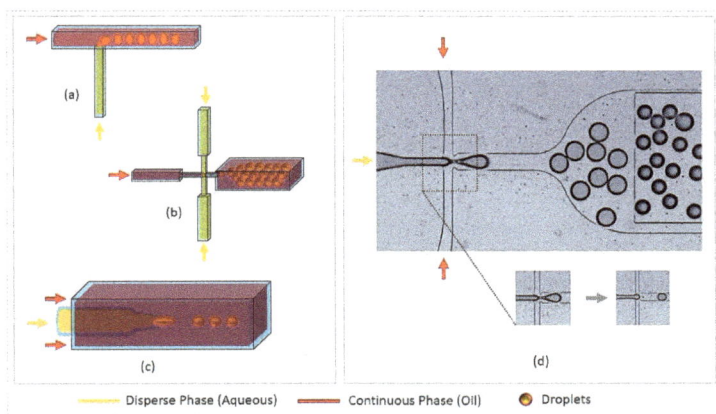

Figure 1. Droplet generation by microfluidic systems. (**a**) T-junction; (**b**) Flow-focusing; (**c**) Co-flow (glass capillary); (**d**) Droplet generation by flow focusing device (use of fluorinated oil with stabilizing agent and disperse phase as water solution of dye), device also includes on-chip storage pool for droplet collection and operation region.

Perfluorinated oils possess two other important features: (i) PDMS-based microfluidic devices swell much less in their presence than in the presence of hydrocarbon oils, (ii) perfluorinated oils absorb large quantities of oxygen and carbon dioxide, which appears to be a very important feature for the encapsulation of cells and other living organisms in droplets. However, particular care should be taken regarding the used surfactant concentration range in order to avoid mass exchange between droplets, which was shown to occur easily at concentrations above the cmc (critical micellar concentration) [109–113]. Many suitable perfluorinated oils (e.g., HFE 7500 or FC 40) are commercially available (3M company, St Paul, MN, USA). Perfluorinated surfactants can either be purchased from RainDance Technologies (Billerica, MA , USA) or Dolomite (Royston, UK), or home-made according to a reaction scheme initially developed by Holtz et al. [105]. The chemical synthesis is based on the condensation of Krytox FSH-157™ (Kry, a perfluoro-polyether (PFPE) carboxylate from DuPont® which acts as an oil-philic moiety) and polyether derivatives like polyethylene glycol (PEG) or Jeffamine® polyetheramine (Huntsman corp.), playing the role of hydrophilic moieties [32,105,107,108]. Commonly used surfactants include the triblock kry-PEG-kry (or PEG-kry2) [105] or the diblock Kry-Jeffa (a contraction of Krytox and Jeffamine®) [107]. One may note that Krytox may also be used as a surfactant in many experiments where biocompatibility is not required. Nevertheless, the negative charge of the Krytox carboxylate group interacts with oppositely charged biomolecules, which may cause the encapsulated biomacromolecules in droplets to lose their activity and aggregate at the droplet interface.

Biocompatible perfluorinated surfactants have also been recently reported, such as LPG-Kry2, where the hydrophilic moiety is a linear poly-glycerol (LPG) derivative [108].

3. Real-Time Fluorescence Measurements of Droplet Size and Size Distribution

For biomedical applications, fluorescence detection is one of the most popular methods. It allows for real-time monitoring of droplet generation rate and droplet analysis. One or more fluorescent probes are generally used. In the case of dual fluorescence acquisition, for instance, one probe is used for the detection of the droplet content or for the detection of a specific biomarker, while the second probe is used for the detection of a second biomarker of interest. A standard dual fluorescence acquisition setup is depicted in Figure 2. It includes two laser sources optimized for the absorption of the two fluorophores. Laser incident beams are combined by means of a first dichroic mirror (DM1) and then directed towards microdroplets in the microfluidic channel by a second dichroic mirror (DM2). The focused band limited light is targeted towards the droplets and recollected by the microscope objective, which is then transmitted through another set of band-limited filters to two photo-multiplier tubes (PMTs). The signal output from PMTs are then collected at high acquisition rates (~100 kHz) using a data acquisition card (DAQ, National Instruments) and analyzed using FPGA (Field-Programmable Gate Array, Labview, National Instruments) module scripts, which allows for the identification of droplets by the modulation of fluorescence versus time.

Figure 2 shows a typical fluorescence intensity real-time recording from droplets flowing in a 30 μm × 30 μm wide microchannel. Each pulse corresponds to the passage of a single droplet, each of which contains two fluorescent dyes (fluorescein and rhodamine in this experiment), which are excited by two continuous wave (CW) lasers at 488 nm and 532 nm. The duration of each pulse corresponds to the residence time (τ) of a single droplet under the illuminated area (lasers footprint) of the microfluidic channel. Measuring the τ value may allow in principle for the determination of the size of the corresponding droplet, provided that the droplet velocity is known. However, if in a single phase flow the mean velocity value, $\bar{u}$, can be easily determined from the flow features—namely the flow rate q and the cross-sectional area S of the microfluidic channel ($\bar{u} = \frac{q}{S}$). This task proves to be cumbersome in the case of a flow laden with deformable microdroplets and more particularly in the presence of large droplets. In this case, the flow is strongly modified due to the formation of a thin lubrication oil film between the droplet interface and the microchannel walls. The presence of such a film has a direct effect on the velocity of the droplets and makes the flow pattern complex and difficult to analyze [114–126]. It has been shown, for instance, that depending on the geometry of the channel, the lubrication film may move either backwards in the case of a cylindrical channel (in regards to a reference frame attached to the droplet) or forward in the case of rectangular or square channels. In the last case, one should take account of the presence in the flow of the continuous oil phase along the gutters of the rectangular or square channels [3,5,127,128].

The determination of the size of droplets becomes straightforward if one accounts for the droplets' generation frequency f and the flow rate of the dispersed phase Q_d, as we demonstrate hereafter in the case of small spherical droplets. Let D_{dr} and $V_{dr} = \frac{\pi}{6}D_{dr}^3$ be their diameter and individual volume, respectively. Since the volume V_{dr} is injected in the microchannel during a period of time $T = \frac{1}{f}$ separating the generation of two successive droplets, one deduces easily D_{dr} according to $D_{dr} = (\frac{6}{\pi}\frac{Q_{dr}}{f})^{\frac{1}{3}}$. We deduce for instance in the case where f = 862 Hz and Q_d = 25 μL/h (results shown in Figure 2), a droplet size D_{dr} = 24.8 μm, which compares very well with the mean droplet size measured directly using droplet image analysis from droplets collected in a dedicated microfluidic observation chamber—that is , 26 μm (results not shown).

Real-time fluorescence measurements can also give a valuable insight into the effect of surface tension value on the droplet size, as illustrated in Figure 3. This figure shows fluorescence intensity recordings versus time from two types of droplets produced with two different values of interfacial tension (all other parameters were kept constant; e.g., flow rate of the dispersed phase was $Q_d = 30$ µL/h). One finds that the droplet size decreased from 31.3 µm to 26.7 µm as the surface tension decreased from $\gamma_1 = 18$ mN/m (Figure 3a) and $\gamma_2 = 13$ mN/m (Figure 3b), respectively, in agreement with a model suggested earlier by Nguyen et al. [129], where the droplet size was shown to vary as the square root of the interfacial tension: $D_{dr} \propto \sqrt{\gamma}$. The observed change in droplet size can be understood in terms of lowering of the energy cost to build the interface between the two non-miscible phases when the interfacial energy decreases.

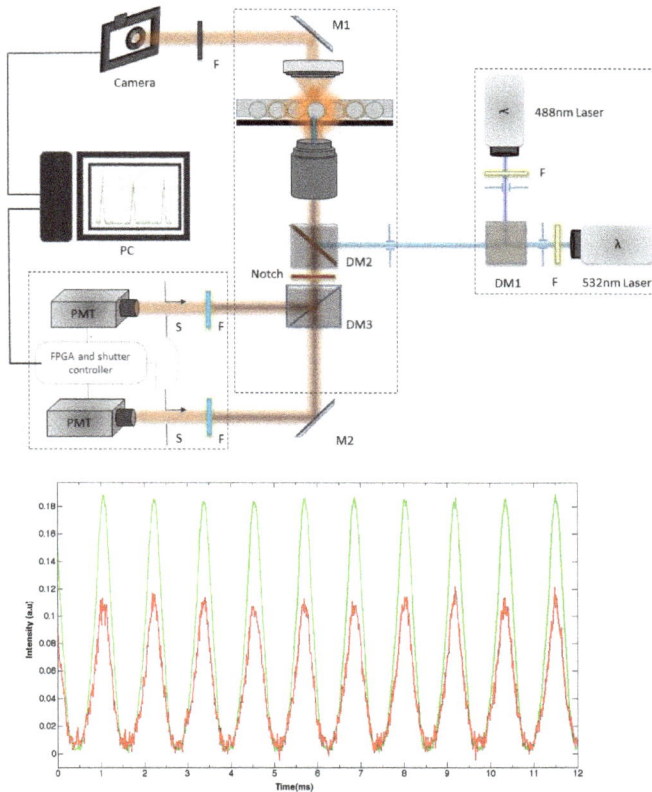

Figure 2. (top) A typical dual channel microfluidic droplet monitoring setup, consisting of two laser sources with each laser band limited by a bandpass filter. The two components of the fluorescence signals emitted by the two different dyes are separated, filtered, and collected on two different photo-multiplier tubes (PMTs). **(bottom)** Typical recorded fluorescence intensity versus time emitted by flowing droplets containing both fluorescein and rhodamine dyes at different concentrations. DM: dichroic mirror.

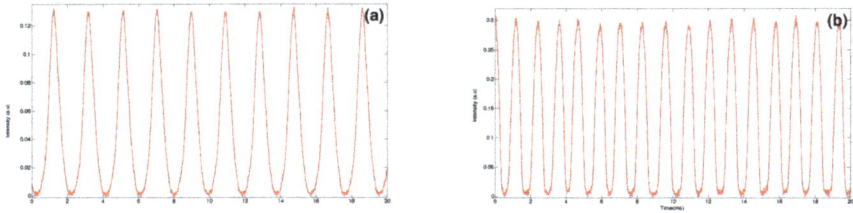

Figure 3. Fluorescence signal extracts from setup (**a**) without surfactant; (**b**) with surfactant.

4. Highly Sensitive Analysis of Droplet Content and Droplet Interface

Real-time fluorescence acquisition not only provides useful information about droplet size and size distribution, it may also give a deeper insight into molecular organization and interactions within the droplet and its interface. For illustration, new results obtained by the authors are presented in Figure 4. They show fluorescence recordings obtained from large droplets containing a rhodamine B fluorescent dye solution (1 mM) and flowing in a roughly square channel with a cross-sectional area of approximately 110 µm × 120 µm. Three different experimental conditions were investigated, differing only in the presence and the type of the used surfactant: Figure 4a corresponds to a case where no surfactant is added; Figure 4b corresponds to a case where a non-ionic surfactant (KryJeffa) is added; Figure 4c,d both correspond to a case where a negatively-charged surfactant (Krytox) was added but for two different droplet sizes—127 µm and 225 µm, respectively.

The bell shape exhibited by the fluorescence intensity of droplets in Figure 4a indicates that droplets more likely adopt a roughly spherical shape: fluorescence intensity starts to increase slightly as the curved interface of the droplet moves more and more across the (still) laser spot before reaching a maximum value when the overlap between the laser footprint and the droplet is at its maximum. One may also note a slight asymmetry of the fluorescence peaks (maximum intensity is shifted towards the left side of the peak). This shift should be correlated to the well-known difference between the profiles of the front and the back of droplets when moving in a flow, as depicted in Figure 4e. It is worth noting that Baret et al. [109] reported direct evidence of the accumulation of surfactant molecules at the back of droplets almost a decade ago, in the flow of a perfluorinated carrier oil (FC40) and using a home-made fluorescent surfactant, namely Krytox-PEO-fluorescein, where a fluorescein isothiocyanate molecule linked to an amine-terminated polyethylene oxide hydrophilic head group was added to Krytox. Interestingly, this surfactant is fluorescent only when the head group is in an aqueous solution, which enabled these authors to monitor the buildup of the surfactant monolayer at the water/oil interface by the readout of droplet interface fluorescence. In contrast, in our study, the accumulation of the surfactant (which is not fluorescent) was indirectly demonstrated by the accumulation of charged rhodamine molecules at the back of the droplets.

In the presence of a surfactant, the recirculation flow induced by the motion of the droplet in the viscous carrier oil generates a heterogeneous distribution of the surfactant at the droplet interface: the surfactant interfacial density becomes higher at the rear region of the droplets than at its front. This effect leads to a rigidification of the droplet interface due to the so-called Marangoni effect [106]. This effect can be clearly seen by comparing fluorescence peaks of Figure 4b–d. In this case, the fluorescence intensity increases rapidly as soon as the laser spot starts to overlap with the droplet front (right region of the fluorescence peaks) and remains almost constant during the flight time τ of the droplet. Hence, in the presence of a surfactant, the droplet deforms less in the flow and is more likely to adopt a plug-like form, as sketched in Figure 4f,g.

In the presence of the negatively-charged Krytox surfactant (Figure 4c,d), one observes a burst of fluorescence at the rear region of the droplet, which corresponds to the left side of the fluorescence peak. By comparing the profiles of the fluorescence peaks obtained with non-ionic surfactant (kryJeffa) and ionic surfactant, the observed burst of fluorescence should be correlated to an increase of the density of the positively-charged fluorescent rhodamine molecules at the rear part of the droplet, which is itself induced by the asymmetric distribution of the surfactant at the droplet interface (see Figure 4f,g).

These results show that the interaction between the droplet content and the surfactant can be detected in a highly sensitive and quantitative manner.

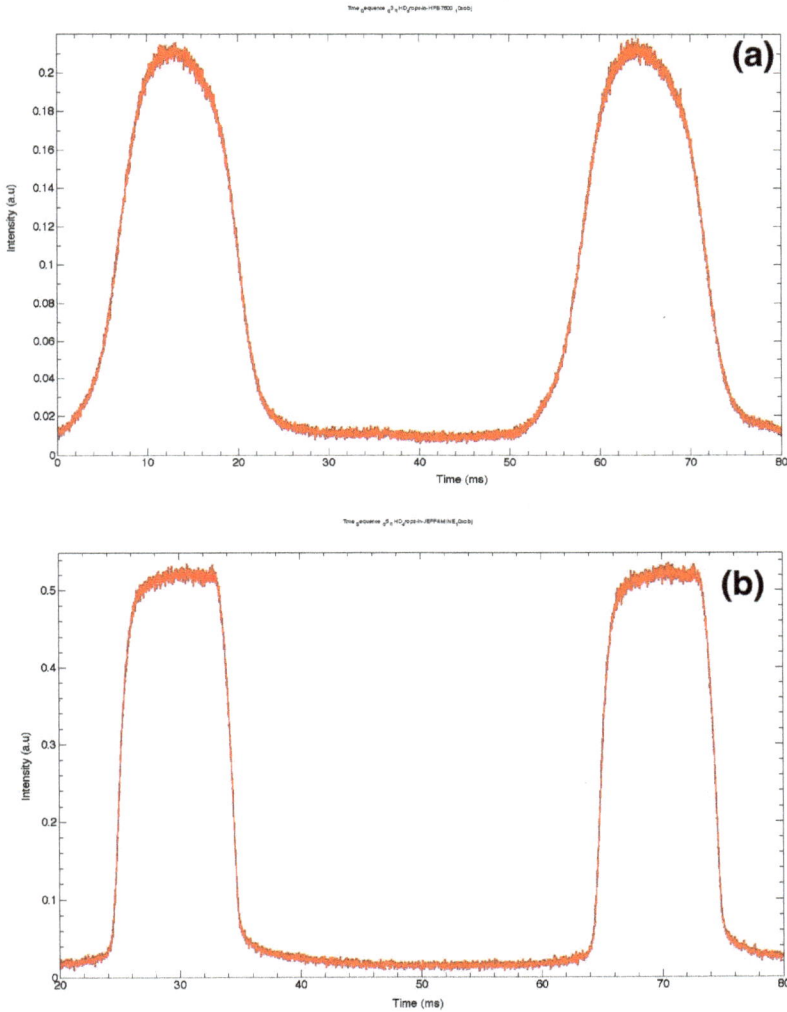

Figure 4. *Cont.*

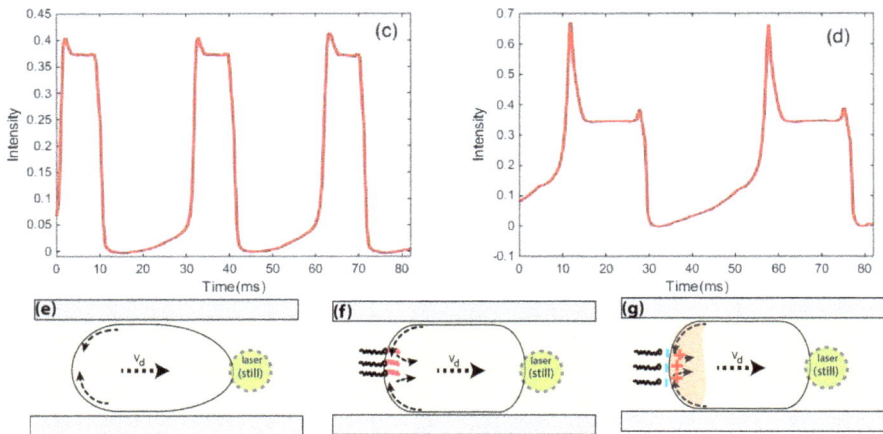

Figure 4. Fluorescence signal from microdroplets (**a**) in HFE7500–without surfactant; (**b**) with surfactant KryJeffa; (**c**) droplets stabilized in Krytox (size around 125 μm); (**d**) big droplets with surfactant Krytox (size above 250 μm); (**e**) plug-like deformation of a large droplet induced by flow of viscous oil; (**f**) heterogeneous distribution of surfactant at the droplet interface; (**g**) in the case of Krytox the distribution of surfactant and its corresponding interaction to charged rhodamine molecules at the rear of the microdroplet.

5. Ultra-High-Throughput Droplets Production and Detection Methods

Droplets may be produced at kHz rates. Nevertheless, a high acquisition of droplet generation frequency includes droplet monitoring and counting operations based on the detection of optical or electrical signals. Reliable counting and sorting play an important role; similarly, the size distribution of the generated population is important for many applications. Several studies have reported the generation of higher-order droplet frequencies, but the state-of-the-art is limited by the acquisition required to monitor in real-time the generation and rate at which the pace is kept. Throughput achieved up to hundreds of thousands of droplets per second could provide in-depth information about analytes but would require hours of active investigation.

To increase both the throughput and the detection limit, many groups have developed improved techniques. Some of the fastest and most reliable methods for ultra-high-throughput utilize, for instance, a laser-induced fluorescence (LIF)-based microfluidic hemocytometer for counting cells encapsulated inside droplets with an average rate of ~600 Hz [130], or amplitude modulation of the acquired signal by a lens-free detection with a rate of 1.7 kHz [33], while a CMOS mounted sensor on a PDMS slab reached a rate of 250 kHz Figure 5C [34]. A high-speed camera assisted by a Fresnel lens reached a read out of 200 kHz [35]. A technique, named "IC 3D" (Integrated Comprehensive Droplet Digital Detection), achieved 100 kHz throughput detection of droplet-encapsulated Blood-DNA, by rotation of a cuvette containing the microdroplets [36] (see system overview in Figure 5B). Another technique where a commercially-available DSLR camera-based system (Figure 5E) developed and recorded at a rate of 250,00 droplets per second [37], while a micro-lens (Figure 5D) array network gathered a read out of 50 kHz [38], and more recently a handy cell phone camera-based read out of one million drops (see Figure 5A) marked the highest data read out and sequencing in digital assays [39].

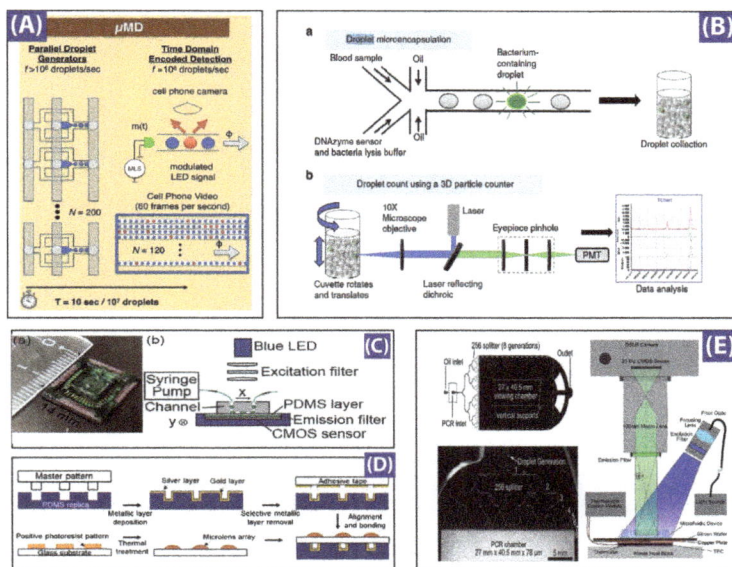

Figure 5. Different studies for droplet generation monitoring: (**A**) A microdroplet megascale detector (μMD) containing a micro-controller-based light emitting diode, a microfluidic device with 120 parallel dropmakers, a cell phone camera for recording, and an off-site data processor. Reproduced with permission from [39]; (**B**) Experimental setup of the Integrated Comprehensive Droplet Digital Detection (IC 3D) system, housing 496 nm and 532 nm laser sources, a dual source single detector scheme modified for the typical experimental needs. A software controlled micro-cuvette holder and rotation unit (1–1100 rpm in rotational speed while 1–15 mm/s vertical translational speed). Reproduced with permission from [36]; (**C**) A CMOS (complementary metal oxide semiconductor)-based sensor with channel bed as closest perimeter for fluorescence detection. The compact sensor assembly consists of a 1280 × 1024 pixel platform, a spin-coated pigment-based band-pass filter, a 250 mW blue LED, and another filter to band limit the light between 457 nm to 492 nm. Reproduced with permission from [34]; (**D**) Integrated micro-optical system with micro-lens assemblies on top of droplet chambers while metallic surfaces at other side of chamber provide optical resonance for improved signal. Reproduced with permission from [38]; (**E**) Experimental stage for digital polymerase chain reaction (PCR) housing a 1 × 256 droplet splitting microfluidic chip on a thermistor stage for PCR thermocycling, a wide field light source, and a digital camera with large field-of-view lens assembly. Reproduced with permission from [37].

Another interesting study, reported recently by Shivhare et al. [47] claims the development of a new cost-effective optofluidic dye-free method allowing for a real-time measure of the mean droplet size of a population of droplets and for the measure of droplet size distribution, which is based on the detection of the forward scattered signal (FSC) of an incident non-focused IR laser beam by flowing droplets in the microchannel and the measure of the residence time of these droplets across the incident laser beam. The used microfluidic device consisted of a main drive channel neighboring two control channels, named as grooves. One groove is dedicated to the input laser signal by means of an optical fiber guide, and the other groove is dedicated to the detection part. When a droplet traverses the detection region, it obstructs the passage of the laser beam, which results in a pulse in the detected signal. Shivhare et al. reported a mean droplet size of 15 μm with approximately 10% discrepancy regarding results obtained from optical image analysis. They postulated that the normalized residence time of the generated droplets along the channel is a better measure of the effective droplet size than forward scattered signal, which is correlated nonlinearly to the droplet size [47]. It is interesting to note that for large droplets, similar to our results shown in Figure 4c,d, Shivhare et al. also report a higher scattered signal from the back of the droplet. This result should also be interpreted as the consequence of a greater rigidification of the back of the droplets induced by the addition of the droplet stabilizing surfactant (Span 85) and the recirculation flow, as a rigid interface should scatter more light than a softer interface.

6. Optically-Assisted Slicing and Merging of Microdroplets

In studies involving bio-molecular assays, the splitting and sorting of microdroplets is a necessity for on-demand droplet size reduction, scale dilution, and volume control of daughter droplets. Splitting can be performed by active methods such as electric field splitting, acoustic, or electro-wetting. Passive splitting could be performed by mechanical in-channel deforming geometries which squeeze and cut the droplet into two. For the passive type, studies have used rectangular and cylindrical channel structures which employ two different types of fabrication modalities. Opto-electrowetting (OEW) [131–139] is a novel technique first reported by Chiou et al. [131], where a photoconductive layer is deposited above the electrodes commonly used in EWOD (electro-wetting over dielectric) [140,141]. The working principle involves a local change in the surface properties (contact angle and surface tension at light-spot) between the conductive layers and the liquid of interest. The use of this kind of OEW device for the optical manipulation of droplets by virtual electrodes [132,134–139], or by local change of the hydrophobicity of the radiated surface provides significant change in the local surface tension; thus, droplets can be moved, merged, patterned, and diffused. The all-optical elemental control involves light-assisted digital microfluidic chips (LADM) [135–138] for droplet movement by a laser source [131–134] or by data projectors [135–139]. Pei et al. [135] use data projection to move, merge, elongate, and divide microdroplets on the surface of a dielectric layer by utilizing a multi-pattern projection method, thus resulting in a unique multiple drop generation, movement, and control (Figure 6a). Later, Pei et al. [136] demonstrated a new approach by introducing a system of Teflon blades (Figure 6b) between two electrodes for slicing microdroplets at the nL scale, and the splitting ratio was reported to be between 10% and 90%.

To add a new study, the breakage of long caged-group molecules grafted on the droplet periphery is a novel idea we reported in a recent work [32]. The technique involves the photolysis of a photosensitive surfactant chain. A pulsed laser source at 1 kHz depletes the surfactant monolayer grafted on the droplet interface. Two surfactant compounds were fabricated with 8-piperazinyl-2-hydroxymethyl-quinoline (8-PHQ), named surfactant 1 and 2 (see Figure 7). Surfactant 1 resulted in stable monodisperse droplet formation and controlled release. Various concentrations resulted in significant reductions in time of merging up to ~2 s, while the second compound (Surfactant 2) was unable to perform the controlled merging of microdroplets. Pendent drop tensiometry (Figure 7b) and Langmuir monolayer method (Figure 7c) were used to evaluate interfacial tension versus concentration and corresponding molecular area occupied by a single molecule under the compression of monolayers. Figure 7d shows the time lapse for the merging of two monodisperse droplets. These methods could develop a new approach for medical diagnostics and treatment for on-the-spot on-demand region-selective target, release, and treat methodology.

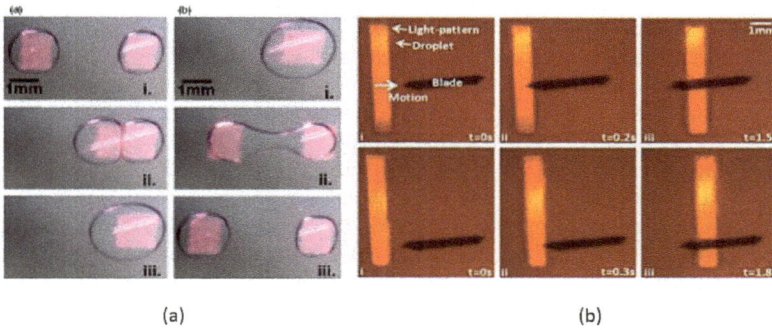

Figure 6. Optical droplets merging and sorting: (**a**) Square-shaped projection patterns moving two droplets towards each other, merged, and sliced back to acquire two droplets. Reproduced with permission from [135]; (**b**) Time sequences of an elongated droplet sliced by on-chip Teflon blade, droplet motion assisted by line projection. Reproduced with permission from [136].

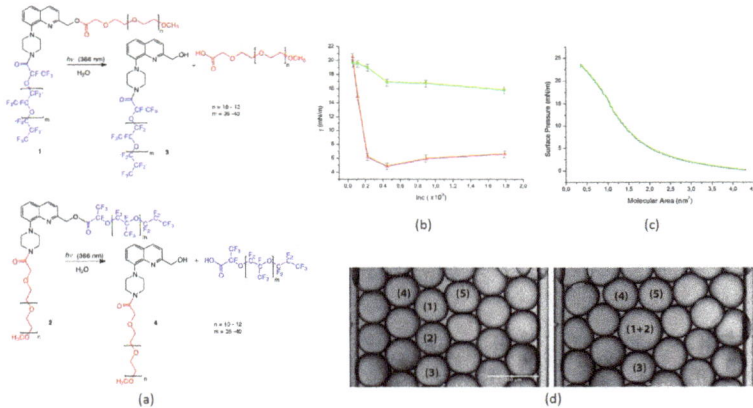

Figure 7. Surfactants 1 and 2: (**a**) Chemical composition; (**b**) Interfacial tension versus concentration, red curve for Surfactant 1 and green curve for Surfactant 2; (**c**) Surface pressure versus molecular area; (**d**) Light-induced merging of droplets 1 and 2. Reproduced with permission from [32].

7. Microdroplets as Optical Probes for 3D Imaging and Sensing

The production of a large population of highly monodisperse droplets with highly controllable optical features gave rise to very interesting applications in the field of 3D imaging and sensing, either as a tunable liquid double emulsion [63], solid spherical particles [55,142], or soft reconfigurable core with elastic shell microparticles [56]. The periphery of the spherical microdroplets performs the convergence or divergence of the incident light provided that the morphology or fabrication protocols are addressed properly. Besides the use of single micro-optical lens systems for imaging and improved resolution, arrayed networks increase spatial resolution and fluorescence signal detection many times over. For instance, Lim et al. [38] reported an eight-times increase in the fluorescence signal by introducing a soft micro-lenses system on the top of metallic coated micro-channels. Additionally, Ghenuche et al. [142] reported on another arrayed micro-optical lenses system based on microspheres, which enables the parallel detection of single fluorescent molecules in a multi-focus nanojet experimental technique. Ghenuche et al. utilized latex micro-spheres of size 2 μm for the

73

generation of such nanojets. In order to perform fluorescence correlation spectroscopy by nanojets, they illuminated a 10 μm spot (approximately 25 micro-spheres) by a low numerical aperture (NA) objective and acquired sensitivity as low as 20 pico-molar fluorescent dye concentration. These type of detection schemes assisted by micro-optical elements yield good estimations of concentration (in the pM range), diffusion coefficient, and relative hydrodynamic radius of the dye molecules.

Another interesting study reported by Nagelberg et al. [63], who used microdroplets as optical microlenses with a tunable focal length based on the concentration variations of the surfactant and the drop–disperse–drop double emulsion phases. They reported the use of hydrocarbons–fluorocarbon–aqueous phases for focal length tunability by adjusting the relationship among the refractive indexes of the interfaces involved, be it the denser fluid inside causing a converging lens or be the denser fluid in the drop shell resulting in a diverging lens system. Figure 8 represents the different types of used droplets, from janus drops to double emulsions and inverse emulsions. The demand for this kind of optical tunability finds potential in super-resolution imaging and displays, light field displays, liquid crystal displays, digital micro-mirror displays, optical tweezers, and medical diagnostic and investigation probes.

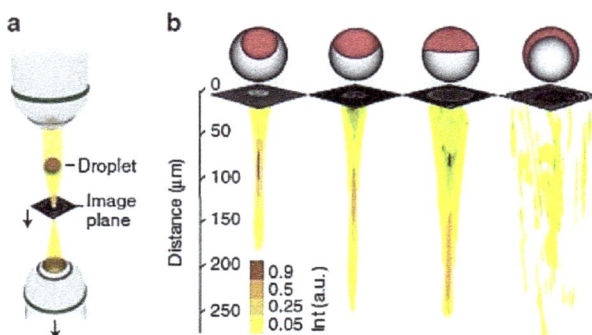

Figure 8. Imaging using microdroplets: setup for light input from the top with an adjustable image plane, droplets on a substrate, image plane, and bottom collection objective with vertical translation, characteristic focal length tunability based on the morphology of the emulsion, and corresponding fully converging-to-fully diverging mechanism. Reproduced with permission from [63].

8. Conclusions

This review outlines recent advances in droplet optofluidics and focuses more particularly on analysis tools for producing highly monodisperse droplets using microfluidic devices and optical methods at high throughput. The optical qualities of micro-droplets and their high potential for applications in biology and chemistry open prospects for applications in drop/capsule/container-on-demand, lab-on-a-chip, cellular matrix mimicking, reconfigurable drug carriers, in-channel processes, and incubation and surface modification in particular as development tools for highly sensitive sensors. We first present the basic concepts of droplet microfluidics, which include microfluidic devices, droplets fabrication, and stabilization. Particular attention is given to experimental optical methods developed for a real-time measurement of droplet size and size distribution, since light provides flexibility and wavelength/intensity tunability. Among the developed optical methods, real-time fluorescence measurement is a highly sensitive one. It allows not only for the detection and analysis of the size and content of droplets, but also for a deep insight into the molecular interaction between droplet contents and the surrounding surfactant molecules. We also present an extension of droplet optofluidics which uses microdroplets as reconfigurable micro-lenses. In summary, droplet microfluidics joined by the potential of optics as a probing and extracting method could open up new dimensions to biomimetic reconstruction and optical control on one hand and smart drug delivery optofluidics micro-systems on the other.

Conflicts of Interest: The authors declare no conflict of interest.

Abbreviations

The following abbreviations are used in this manuscript:

DNA	Deoxyribonucleic acid
SAXS	small angle X-ray scattering
PDMS	Polydimethylsiloxane
ILIDS	interferometric laser imaging for droplet sizing
CMOS	Complementary metal–oxide–semiconductor
IC 3D	integrated comprehensive droplet digital detection
DSLR	Digital single-lens reflex camera
μMD	Microdroplet megascale detector
PCR	Polymerase chain reaction
LIF	Laser-induced fluorescence
PEG	Polyethylene glycol
EWOD	Electrowetting over dielectric
OEW	Opto-electrowetting
PS	Polystyrene
PMMA	Poly-methyl-methacrylate
PC	Polycarbonate
FSC	Forward scattered signal
LADM	Light assisted digital microfluidic chip
W/O	Water-in-oil emulsion
O/W	Oil-in-water emulsion
pH	Potential of hydrogen

References

1. Thorsen, T.; Robert, R.W.; Arnold, F.H.; Quake, S.R. Dynamic Pattern Formation in a Vesicle-Generating Microfluidic Device. *Phys. Rev. Lett.* **2001**, *86*, 4163–4166. [CrossRef]
2. Thorsen, T.; Maerkl, S.J.; Quake, S.R. Microfluidic Large Scale Integration. *Science* **2002**, *298*, 580–584. [CrossRef]
3. Baroud, C.N.; Gallaire, F.; Dangla, R. Dynamics of microfluidic droplets. *Lab Chip* **2010**, *10*, 2032–2045.
4. Gu, H.; Duits, M.H.G.; Mugele, F. Droplets Formation and Merging in Two-Phase Flow Microfluidics. *Int. J. Mol. Sci.* **2011**, *12*, 2572–2597.
5. Seemann, R.; Brinkmann, M.; Pfohl, T.; Herminghaus, S. Droplet based microfluidics. *Rep. Prog. Phys.* **2012**, *75*, 16601–16642.
6. Agresti, J.J.; Antipov, E.; Abate, A.R.; Ahn, K.; Rowat, A.C.; Baret, J.C.; Marquez, M.; Klibanov, A.M.; Griffiths, A.D.; Weitz, D.A. Ultra-high-throughput screening in drop-based microfluidics for directed evolution. *Proc. Natl. Acad. Sci. USA* **2010**, *107*, 4004–4009.
7. deMello, A.J. Control and detection of chemical reactions in microfluidic systems. *Nature* **2006**, *442*, 394–402.
8. Song, H.; Chen, D.L.; Ismagilov, R.F. Reactions in Droplets in Microfluidic Channels. *Angew. Chem. Int. Ed.* **2006**, *45*, 7336–7356.
9. Witters, D.; Sun, B.; Begolo, S.; Rodriguez-Manzano, J.; Robles, W.; Ismagilov, R.F. Digital biology and chemistry. *Lab Chip* **2014**, *14*, 3225–3232.
10. Kumacheva, E.; Garstecki, P. *Microfluidic Reactors for Polymer Particles*; Wiley: Chichester, UK, 2011.
11. Xi, H.D.; Zheng, H.; Guo, W.; Ganan-Calvo, A.M.; Ai, Y.; Tsao, C.W.; Zhou, J.; Li, W.; Huang, Y.; Nguyen, N.T.; et al. Active droplet sorting in microfluidics: A review. *Lab Chip* **2017**, *17*, 751–771.
12. Taly, V.; Pekin, D.; El Abed, A.; Laurent-Puig, P. Detecting biomarkers with microdroplet technology. *Trends Mol. Med.* **2012**, *18*, 405–416.

13. Pekin, D.; Skhiri, Y.; Baret, J.C.; Le Corre, D.; Mazutis, L.; Ben Salem, C.; Millot, F.; El Harrak, A.; Hutchison, J.B.; Larson, J.W.; et al. Quantitative and sensitive detection of rare mutations using droplet-based microfluidics. *Lab Chip* **2011**, *11*, 2156–2166.

14. Abate, A.R.; Weitz, D.A. Syringe-vacuum microfluidics: A portable technique to create monodisperse emulsions. *Biomicrofluidics* **2011**, *5*, 014107, doi:10.1063/1.3567093.

15. Tan, Y.C.; Hettiarachchi, K.; Siu, M.; Pan, Y.R.; Lee, A.P. Controlled microfluidic encapsulation of cells, proteins, and microbeads in lipid vesicles. *J. Am. Chem. Soc.* **2006**, *128*, 5656–5658.

16. Khan, I.U.; Serra, C.A.; Anton, N.; Vandamme, T. Microfluidics: A focus on improved cancer targeted drug delivery systems. *J. Controll. Release* **2013**, *172*, 1065–1074.

17. Hong, J.; Edel, J.B. Micro-and nanofluidic systems for high-throughput biological screening. *Drug Discov. Today* **2009**, *14*, 134–146.

18. Christopher, G.F.; Anna, S.L. Microfluidic methods for generating continuous droplet streams. *J. Phys. D Appl. Phys.* **2007**, *40*, R319.

19. Chin, L.K.; Lee, C.H.; Chen, B.C. Imaging live cells at high spatiotemporal resolution for lab-on-a-chip applications. *Lab Chip* **2016**, *16*, 2014–2024.

20. Lau, A.K.S.; Shum, H.C.; Wong, K.K.Y.; Tsia, K.K. Optofluidic time-stretch imaging—An emerging tool for high-throughput imaging flow cytometry. *Lab Chip* **2016**, *16*, 1743–1756.

21. Friedrich, S.M.; Zec, H.C.; Wang, T.H. Analysis of single nucleic acid molecules in micro- and nano-fluidics. *Lab Chip* **2016**, *16*, 790–811.

22. Liu, P.Y.; Chin, L.K.; Ser, W.; Chen, H.F.; Hsieh, C.M.; Lee, C.H.; Sung, K.B.; Ayi, T.C.; Yap, P.H.; Liedberg, B.; et al. Cell refractive index for cell biology and disease diagnosis: Past, present and future. *Lab Chip* **2016**, *16*, 634–644.

23. Zhao, H.T.; Yang, Y.; Chin, L.K.; Chen, H.F.; Zhu, W.M.; Zhang, J.B.; Yap, P.H.; Liedberg, B.; Wang, K.; Wang, G.; et al. Optofluidic lens with low spherical and low field curvature aberrations. *Lab Chip* **2016**, *16*, 1617–1624.

24. Fan, S.K.; Lee, H.P.; Chien, C.C.; Lu, Y.W.; Chiu, Y.; Lin, F.Y. Reconfigurable liquid-core/liquid-cladding optical waveguides with dielectrophoresis-driven virtual microchannels on an electromicrofluidic platform. *Lab Chip* **2016**, *16*, 847–854.

25. Fan, S.K.; Wang, F.M. Multiphase optofluidics on an electro-microfluidic platform powered by electrowetting and dielectrophoresis. *Lab Chip* **2014**, *14*, 2728–2738.

26. Shui, L.; Hayes, R.A.; Jin, M.; Zhang, X.; Bai, P.; van den Berg, A.; Zhou, G. Microfluidics for electronic paper-like displays. *Lab Chip* **2014**, *14*, 2374–2384.

27. Huang, N.T.; Zhang, H.L.; Chung, M.T.; Seo, J.H.; Kurabayashi, K. Recent advancements in optofluidics-based single-cell analysis: Optical on-chip cellular manipulation, treatment, and property detection. *Lab Chip* **2014**, *14*, 1230–1245.

28. Wang, N.; Zhang, X.; Wang, Y.; Yu, W.; Chan, H.L.W. Microfluidic reactors for photocatalytic water purification. *Lab Chip* **2014**, *14*, 1074–1082.

29. Yu, J.Q.; Huang, W.; Chin, L.K.; Lei, L.; Lin, Z.P.; Ser, W.; Chen, H.; Ayi, T.C.; Yap, P.H.; Chen, C.H.; et al. Droplet optofluidic imaging for [small lambda]-bacteriophage detection via co-culture with host cell *Escherichia coli*. *Lab Chip* **2014**, *14*, 3519–3524.

30. Gaber, N.; Malak, M.; Marty, F.; Angelescu, D.E.; Richalot, E.; Bourouina, T. Optical trapping and binding of particles in an optofluidic stable Fabry-Perot resonator with single-sided injection. *Lab Chip* **2014**, *14*, 2259–2265.

31. Muller, P.; Kopp, D.; Llobera, A.; Zappe, H. Optofluidic router based on tunable liquid-liquid mirrors. *Lab Chip* **2014**, *14*, 737–743.

32. Dunkel, P.; Hayat, Z.; Barosi, A.; Bchellaoui, N.; Dhimane, H.; Dalko, P.I.; El Abed, A.I. Photolysis-driven merging of microdroplets in microfluidic chambers. *Lab Chip* **2016**, *16*, 1484–1491.

33. Muluneh, M.; Kim, B.; Buchsbaum, G.; Issadore, D. Miniaturized, multiplexed readout of droplet-based microfluidic assays using time-domain modulation. *Lab Chip* **2014**, *14*, 4638–4646.

34. Kim, M.; Pan, M.; Gai, Y.; Pang, S.; Han, C.; Yang, C.; Tang, S.K. Optofluidic ultrahigh-throughput detection of fluorescent drops. *Lab Chip* **2015**, *15*, 1417–1423.

35. Schonbrun, E.; Abate, A.R.; Steinvurzel, P.E.; Weitz, D.A.; Crozier, K.B. High-throughput fluorescence detection using an integrated zone- plate array. *Lab Chip* **2010**, *10*, 852–856.

36. Kang, D.K.; Ali, M.M.; Zhang, K.; Huang, S.S.; Peterson, E.; Digman, M.A.; Gratton, E.; Zhao, W. Rapid detection of single bacteria in unprocessed blood using Integrated Comprehensive Droplet Digital Detection. *Nat. Commun.* **2014**, *5*, doi:10.1038/ncomms6427.

37. Hatch, A.C.; Fisher, J.S.; Tovar, A.R.; Hsieh, A.T.; Lin, R.; Pentoney, S.L.; Yang, D.L.; Lee, A.P. 1-Million droplet array with wide-field fluorescence imaging for digital PCR. *Lab Chip* **2011**, *11*, 3838–3845.

38. Lim, J.; Gruner, P.; Konrad, M.; Baret, J.C. Micro-optical lens array for fluorescence detection in droplet-based microfluidics. *Lab Chip* **2013**, *13*, 1472–1475.

39. Yelleswarapu, V.R.; Jeong, H.H.; Yadavali, S.; Issadore, D. Ultra-high throughput detection (1 million droplets per second) of fluorescent droplets using a cell phone camera and time domain encoded optofluidics. *Lab Chip* **2017**, *17*, 1083–1094.

40. Dalgleish, D.; Hallett, F. Dynamic light scattering: Applications to food systems. *Food Res. Int.* **1995**, *28*, 181–193.

41. Aichele, C.P.; Venkataramani, D.; Smay, J.E.; McCann, M.H.; Richter, S.; Khanzadeh-Moradllo, M.; Aboustait, M.; Ley, M.T. A comparison of automated scanning electron microscopy (ASEM) and acoustic attenuation spectroscopy (AAS) instruments for particle sizing. *Colloids Surf. A Physicochem. Eng. Aspects* **2015**, *479*, 46–51.

42. Dukhin, A.S.; Goetz, P.J. Acoustic and electroacoustic spectroscopy. *Langmuir* **1996**, *12*, 4336–4344.

43. Miller, C.; Sudol, E.; Silebi, C.; El-Aasser, M. Capillary hydrodynamic fractionation (CHDF) as a tool for monitoring the evolution of the particle size distribution during miniemulsion polymerization. *J. Colloid Interface Sci.* **1995**, *172*, 249–256.

44. Krebs, T.; Ershov, D.; Schroen, C.; Boom, R. Coalescence and compression in centrifuged emulsions studied with in situ optical microscopy. *Soft Matter* **2013**, *9*, 4026–4035.

45. Caporaso, N.; Genovese, A.; Burke, R.; Barry-Ryan, C.; Sacchi, R. Effect of olive mill wastewater phenolic extract, whey protein isolate and xanthan gum on the behaviour of olive O/W emulsions using response surface methodology. *Food Hydrocolloids* **2016**, *61*, 66–76.

46. Kunstmann-Olsen, C.; Hanczyc, M.M.; Hoyland, J.; Rasmussen, S.; Rubahn, H.G. Uniform droplet splitting and detection using lab-on-chip flow cytometry on a microfluidic PDMS device. *Sens. Actuators B Chem.* **2016**, *229*, 7–13.

47. Shivhare, P.; Prabhakar, A.; Sen, A. Optofluidics based lab-on-chip device for in situ measurement of mean droplet size and droplet size distribution of an emulsion. *J. Micromech. Microeng.* **2017**, *27*, 035003, doi:10.1088/1361-6439/aa53cc.

48. Bachalo, W.D. Method for measuring the size and velocity of spheres by dual-beam light-scatter interferometry. *Appl. Opt.* **1980**, *19*, 363–370.

49. Glover, A.; Skippon, S.; Boyle, R. Interferometric laser imaging for droplet sizing: A method for droplet-size measurement in sparse spray systems. *Appl. Opt.* **1995**, *34*, 8409–8421.

50. Querel, A.; Lemaitre, P.; Brunel, M.; Porcheron, E.; Gréhan, G. Real-time global interferometric laser imaging for the droplet sizing (ILIDS) algorithm for airborne research. *Meas. Sci. Technol.* **2009**, *21*, 015306, doi:10.1088/0957-0233/21/1/015306.

51. Shen, H.; Coëtmellec, S.; Gréhan, G.; Brunel, M. Interferometric laser imaging for droplet sizing revisited: Elaboration of transfer matrix models for the description of complete systems. *Appl. Opt.* **2012**, *51*, 5357–5368.

52. Theodorou, E.; Scanga, R.; Twardowski, M.; Snyder, M.P.; Brouzes, E. A Droplet Microfluidics Based Platform for Mining Metagenomic Libraries for Natural Compounds. *Micromachines* **2017**, *8*, 230, doi:10.3390/mi8080230.

53. Czekalska, M.A.; Kaminski, T.S.; Horka, M.; Jakiela, S.; Garstecki, P. An Automated Microfluidic System for the Generation of Droplet Interface Bilayer Networks. *Micromachines* **2017**, *8*, 93, doi:10.3390/mi8030093.

54. Rodríguez-Ruiz, I.; Radajewski, D.; Charton, S.; Phamvan, N.; Brennich, M.; Pernot, P.; Bonneté, F.; Teychené, S. Innovative High-Throughput SAXS Methodologies Based on Photonic Lab-on-a-Chip Sensors: Application to Macromolecular Studies. *Sensors* **2017**, *17*, 1266, doi:10.3390/s17061266.

55. Bchellaoui, N.; Hayat, Z.; Mami, M.; Dorbez-Sridi, R.; El Abed, A.I. Microfluidic-assisted Formation of Highly Monodisperse and Mesoporous Silica Soft Microcapsules. *Sci. Rep.* **2017**, *7*, 16326, doi:10.1038/s41598-017-16554-4.

56. Darafsheh, A.; Guardiola, C.; Palovcak, A.; Finlay, J.C.; Cárabe, A. Optical super-resolution imaging by high-index microspheres embedded in elastomers. *Opt. Lett.* **2015**, *40*, 5–8.

57. Chin, L.; Liu, A.; Zhang, J.; Lim, C.; Soh, Y. An on-chip liquid tunable grating using multiphase droplet microfluidics. *Appl. Phys. Lett.* **2008**, *93*, 164107, doi:10.1063/1.3009560.

58. Yu, J.; Yang, Y.; Liu, A.; Chin, L.; Zhang, X. Microfluidic droplet grating for reconfigurable optical diffraction. *Opt. Lett.* **2010**, *35*, 1890–1892.

59. Chin, L.; Liu, A.; Soh, Y.; Lim, C.; Lin, C. A reconfigurable optofluidic Michelson interferometer using tunable droplet grating. *Lab Chip* **2010**, *10*, 1072–1078.

60. Shen, Z.; Zou, Y.; Chen, X. Characterization of microdroplets using optofluidic signals. *Lab Chip* **2012**, *12*, 3816–3820.

61. Paulsen, K.S.; Di Carlo, D.; Chung, A.J. Optofluidic fabrication for 3D-shaped particles. *Nat. Commun.* **2015**, *6*, doi:10.1038/ncomms7976.

62. Paulsen, K.S.; Chung, A.J. Non-spherical particle generation from 4D optofluidic fabrication. *Lab Chip* **2016**, *16*, 2987–2995.

63. Nagelberg, S.; Zarzar, L.D.; Nicolas, N.; Subramanian, K.; Kalow, J.A.; Sresht, V.; Blankschtein, D.; Barbastathis, G.; Kreysing, M.; Swager, T.M.; et al. Reconfigurable and responsive droplet-based compound micro-lenses. *Nat. Commun.* **2017**, *8*, doi:10.1038/ncomms14673.

64. Zantow, M.; Dendere, R.; Douglas, T.S. Image-based analysis of droplets in microfluidics. In Proceedings of the 2013 35th Annual International Conference of the IEEE Engineering in Medicine and Biology Society (EMBC), Osaka, Japan, 3–7 July 2013; pp. 1776–1779.

65. Girault, M.; Kim, H.; Arakawa, H.; Matsuura, K.; Odaka, M.; Hattori, A.; Terazono, H.; Yasuda, K. An on-chip imaging droplet-sorting system: A real-time shape recognition method to screen target cells in droplets with single cell resolution. *Sci. Rep.* **2017**, *7*, 40072, doi:10.1038/srep40072.

66. Zang, E.; Brandes, S.; Tovar, M.; Martin, K.; Mech, F.; Horbert, P.; Henkel, T.; Figge, M.T.; Roth, M. Real-time image processing for label-free enrichment of Actinobacteria cultivated in picolitre droplets. *Lab Chip* **2013**, *13*, 3707–3713.

67. Kim, H.; Terazono, H.; Nakamura, Y.; Sakai, K.; Hattori, A.; Odaka, M.; Girault, M.; Arao, T.; Nishio, K.; Miyagi, Y.; et al. Development of on-chip multi-imaging flow cytometry for identification of imaging biomarkers of clustered circulating tumor cells. *PLoS ONE* **2014**, *9*, e104372.

68. Reyes, D.R.; Iossifidis, D.; Auroux, P.A.; Manz, A. Micro total analysis systems. 1. Introduction, theory, and technology. *Anal. Chem.* **2002**, *74*, 2623–2636.

69. Squires, T.M.; Quake, S.R. Microfluidics: Fluid physics at the nanoliter scale. *Rev. Mod. Phys.* **2005**, *77*, 977.

70. Sollier, E.; Murray, C.; Maoddi, P.; Di Carlo, D. Rapid prototyping polymers for microfluidic devices and high pressure injections. *Lab Chip* **2011**, *11*, 3752–3765.

71. Berthier, E.; Young, E.W.; Beebe, D. Engineers are from PDMS-land, Biologists are from Polystyrenia. *Lab Chip* **2012**, *12*, 1224–1237.

72. Huang, G.Y.; Zhou, L.H.; Zhang, Q.C.; Chen, Y.M.; Sun, W.; Xu, F.; Lu, T.J. Microfluidic hydrogels for tissue engineering. *Biofabrication* **2011**, *3*, 012001, doi:10.1088/1758-5082/3/1/012001.

73. Khademhosseini, A.; Vacanti, J.P.; Langer, R. Progress in tissue engineering. *Sci. Am.* **2009**, *300*, 64–71.

74. Wu, Z.; Chen, H.; Liu, X.; Zhang, Y.; Li, D.; Huang, H. Protein adsorption on poly (*N*-vinylpyrrolidone)-modified silicon surfaces prepared by surface-initiated atom transfer radical polymerization. *Langmuir* **2009**, *25*, 2900–2906.

75. Pan, T.; Fiorini, G.S.; Chiu, D.T.; Woolley, A.T. In-channel atom-transfer radical polymerization of thermoset polyester microfluidic devices for bioanalytical applications. *Electrophoresis* **2007**, *28*, 2904–2911.

76. Wang, Y.; Chen, H.; He, Q.; Soper, S.A. A high-performance polycarbonate electrophoresis microchip with integrated three-electrode system for end-channel amperometric detection. *Electrophoresis* **2008**, *29*, 1881–1888.

77. Zhou, J.; Ren, K.; Zheng, Y.; Su, J.; Zhao, Y.; Ryan, D.; Wu, H. Fabrication of a microfluidic Ag/AgCl reference electrode and its application for portable and disposable electrochemical microchips. *Electrophoresis* **2010**, *31*, 3083–3089.

78. Zhang, W.; Lin, S.; Wang, C.; Hu, J.; Li, C.; Zhuang, Z.; Zhou, Y.; Mathies, R.A.; Yang, C.J. PMMA/PDMS valves and pumps for disposable microfluidics. *Lab Chip* **2009**, *9*, 3088–3094.

79. Chen, Y.; Zhang, L.; Chen, G. Fabrication, modification, and application of poly (methyl methacrylate) microfluidic chips. *Electrophoresis* **2008**, *29*, 1801–1814.

80. Yang, W.; Yu, M.; Sun, X.; Woolley, A.T. Microdevices integrating affinity columns and capillary electrophoresis for multibiomarker analysis in human serum. *Lab Chip* **2010**, *10*, 2527–2533.

81. Klasner, S.A.; Metto, E.C.; Roman, G.T.; Culbertson, C.T. Synthesis and characterization of a poly (dimethylsiloxane)-poly(ethylene oxide) block copolymer for fabrication of amphiphilic surfaces on microfluidic devices. *Langmuir* **2009**, *25*, 10390–10396.

82. Xia, Y.; Whitesides, G.M. Soft Lithography. *Angew. Chem. Int. Ed.* **1998**, *37*, 550–575.

83. Qin, D.; Xia, Y.; Whitesides, G.M. Soft lithography for micro-and nanoscale patterning. *Nat. Protoc.* **2010**, *5*, 491.

84. Garstecki, P.; Fuerstman, M.J.; Stone, H.A.; Whitesides, G.M. Formation of droplets and bubbles in a microfluidic T-junction—Scaling and mechanism of break-up. *Lab Chip* **2006**, *6*, 437–446.

85. Nisisako, T.; Torii, T.; Higuchi, T. Droplet formation in a microchannel network. *Lab Chip* **2002**, *2*, 24–26.

86. Xu, J.; Li, S.; Tan, J.; Luo, G. Correlations of droplet formation in T-junction microfluidic devices: From squeezing to dripping. *Microfluid. Nanofluid.* **2008**, *5*, 711–717.

87. Chakraborty, I.; Ricouvier, J.; Yazhgur, P.; Tabeling, P.; Leshansky, A. Modeling of droplet generation at shallow microfluidic T-junction. In Proceedings of the Bulletin of the American Physical Society APS March Meeting, Los Angeles, CA, USA, 5–9 March 2018.

88. Garstecki, P.; Gitlin, I.; DiLuzio, W.; Whitesides, G.M.; Kumacheva, E.; Stone, H.A. Formation of monodisperse bubbles in a microfluidic flow-focusing device. *Appl. Phys. Lett.* **2004**, *85*, 2649–2651.

89. Dixon, A.J.; Rickel, J.M.R.; Shin, B.D.; Klibanov, A.L.; Hossack, J.A. In Vitro Sonothrombolysis Enhancement by Transiently Stable Microbubbles Produced by a Flow-Focusing Microfluidic Device. *Ann. Biomed. Eng.* **2018**, *46*, 222–232.

90. Mu, K.; Si, T.; Li, E.; Xu, R.X.; Ding, H. Numerical study on droplet generation in axisymmetric flow focusing upon actuation. *Phys. Fluids* **2018**, *30*, 012111, doi:10.1063/1.5009601.

91. Xu, S.; Nie, Z.; Seo, M.; Lewis, P.; Kumacheva, E.; Stone, H.A.; Garstecki, P.; Weibel, D.B.; Gitlin, I.; Whitesides, G.M. Generation of monodisperse particles by using microfluidics: Control over size, shape, and composition. *Angew. Chem.* **2005**, *117*, 734–738.

92. Nisisako, T.; Torii, T.; Takahashi, T.; Takizawa, Y. Synthesis of monodisperse bicolored janus particles with electrical anisotropy using a microfluidic Co-Flow system. *Adv. Mater.* **2006**, *18*, 1152–1156.

93. Tumarkin, E.; Kumacheva, E. Microfluidic generation of microgels from synthetic and natural polymers. *Chem. Soc. Rev.* **2009**, *38*, 2161–2168.

94. Shah, R.K.; Shum, H.C.; Rowat, A.C.; Lee, D.; Agresti, J.J.; Utada, A.S.; Chu, L.Y.; Kim, J.W.; Fernandez-Nieves, A.; Martinez, C.J.; et al. Designer emulsions using microfluidics. *Mater. Today* **2008**, *11*, 18–27.

95. Bonat Celli, G.; Abbaspourrad, A. Tailoring Delivery System Functionality Using Microfluidics. *Ann. Rev. Food Sci. Technol.* **2018**, *9*, 481–501.

96. Duncanson, W.J.; Lin, T.; Abate, A.R.; Seiffert, S.; Shah, R.K.; Weitz, D.A. Microfluidic synthesis of advanced microparticles for encapsulation and controlled release. *Lab Chip* **2012**, *12*, 2135–2145.

97. Liu, E.Y.; Jung, S.; Weitz, D.A.; Yi, H.; Choi, C.H. High-throughput double emulsion-based microfluidic production of hydrogel microspheres with tunable chemical functionalities toward biomolecular conjugation. *Lab Chip* **2018**, doi:10.1039/C7LC01088E.

98. Ferraro, D.; Champ, J.; Teste, B.; Serra, M.; Malaquin, L.; Viovy, J.L.; De Cremoux, P.; Descroix, S. Microfluidic platform combining droplets and magnetic tweezers: Application to HER2 expression in cancer diagnosis. *Sci. Rep.* **2016**, *6*, 25540, doi:10.1038/srep25540.

99. Ainla, A.; Jansson, E.T.; Stepanyants, N.; Orwar, O.; Jesorka, A. A microfluidic pipette for single-cell pharmacology. *Anal. Chem.* **2010**, *82*, 4529–4536.

100. Zhu, P.; Wang, L. Passive and active droplet generation with microfluidics: A review. *Lab Chip* **2017**, *17*, 34–75.

101. Willaime, H.; Barbier, V.; Kloul, L.; Maine, S.; Tabeling, P. Arnold Tongues in a Microfluidic Drop Emitter. *Phys. Rev. Lett.* **2006**, *96*, 054501, doi:10.1103/PhysRevLett.96.054501.

102. Utada, A.S.; Fernandez-Nieves, A.; Stone, H.A.; Weitz, D.A. Dripping to Jetting Transitions in Coflowing Liquid Streams. *Phys. Rev. Lett.* **2007**, *99*, 094502, doi:10.1103/PhysRevLett.99.094502.

103. Guillot, P.; Colin, A.; Utada, A.S.; Ajdari, A. Stability of a Jet in Confined Pressure-Driven Biphasic Flows at Low Reynolds Numbers. *Phys. Rev. Lett.* **2007**, *99*, 104502, doi:10.1103/PhysRevLett.99.104502.

104. Herrada, M.A.; Gañán Calvo, A.M.; Guillot, P. Spatiotemporal instability of a confined capillary jet. *Phys. Rev. E* **2008**, *78*, 046312, doi:10.1103/PhysRevE.78.046312.

105. Holtze, C.; Rowat, A.C.; Agresti, J.J.; Hutchison, J.B.; Angile, F.E.; Schmitz, C.H.J.; Koster, S.; Duan, H.; Humphry, K.J.; Scanga, R.A.; et al. Biocompatible surfactants for water-in-fluorocarbon emulsions. *Lab Chip* **2008**, *8*, 1632–1639.

106. Baret, J.C. Surfactants in droplet-based microfluidics. *Lab Chip* **2012**, *8*, 422–433.

107. Chiu, Y.L.; Chan, H.F.; Phua, K.K.L.; Zhang, Y.; Juul, S.; Knudsen, B.R.; Ho, Y.P.; Leong, K.W. Synthesis of Fluorosurfactants for Emulsion-Based Biological Applications. *ACS Nano* **2014**, *8*, 3913–3920,

108. Wagner, O.; Thiele, J.; Weinhart, M.; Mazutis, L.; Weitz, D.A.; Huck, W.T.S.; Haag, R. Biocompatible fluorinated polyglycerols for droplet microfluidics as an alternative to PEG-based copolymer surfactants. *Lab Chip* **2016**, *16*, 65–69.

109. Baret, J.C.; Kleinschmidt, F.; El Harrak, A.; Griffiths, A.D. Kinetic Aspects of Emulsion Stabilization by Surfactants: A Microfluidic Analysis. *Langmuir* **2009**, *25*, 6088–6093.

110. Skhiri, Y.; Gruner, P.; Semin, B.; Brosseau, Q.; Pekin, D.; Mazutis, L.; Goust, V.; Kleinschmidt, F.; El Harrak, A.; Hutchison, J.B.; et al. Dynamics of molecular transport by surfactants in emulsions. *Soft Matter* **2012**, *8*, 10618–10627.

111. Fallah-Araghi, A.; Meguellati, K.; Baret, J.C.; Harrak, A.E.; Mangeat, T.; Karplus, M.; Ladame, S.; Marques, C.M.; Griffiths, A.D. Enhanced Chemical Synthesis at Soft Interfaces: A Universal Reaction-Adsorption Mechanism in Microcompartments. *Phys. Rev. Lett.* **2014**, *112*, 028301, doi:10.1103/PhysRevLett.112.028301.

112. Gruner, P.; Riechers, B.; Orellana, L.A.C.; Brosseau, Q.; Maes, F.; Beneyton, T.; Pekin, D.; Baret, J.C. Stabilisers for water-in-fluorinated-oil dispersions: Key properties for microfluidic applications. *Curr. Opin. Colloid Interface Sci.* **2015**, *20*, 183–191.

113. Gruner, P.; Riechers, B.; Semin, B.; Lim, J.; Johnston, A.; Short, K.; Baret, J.C. Controlling molecular transport in minimal emulsions. *Nat.Commun.* **2016**, *7*, doi:10.1038/ncomms10392.

114. Fairbrother, F.; Stubbs, A.E. 119. Studies in electro-endosmosis. Part VI. The "bubble-tube" method of measurement. *J. Chem. Soc.* **1935**, 527–529, doi:10.1039/JR9350000527.

115. Taylor, G. Deposition of a viscous fluid on the wall of a tube. *J. Fluid Mech.* **1961**, *10*, 161–165.

116. Bretherton, F. The motion of long bubbles in tubes. *J. Fluid Mech.* **1961**, *10*, 166–188.

117. Ratulowski, J.; Chang, H.C. Transport of gas bubbles in capillaries. *Phys. Fluids A Fluid Dyn.* **1989**, *1*, 1642–1655.

118. Hodges, S.; Jensen, O.; Rallison, J. The motion of a viscous drop through a cylindrical tube. *J. Fluid Mech.* **2004**, *501*, 279–301.

119. Wong, H.; Radke, C.; Morris, S. The motion of long bubbles in polygonal capillaries. Part 1. Thin films. *J. Fluid Mech.* **1995**, *292*, 71–94.

120. Wong, H.; Radke, C.; Morris, S. The motion of long bubbles in polygonal capillaries. Part 2. Drag, fluid pressure and fluid flow. *J. Fluid Mech.* **1995**, *292*, 95–110.

121. Schwartz, L.; Princen, H.; Kiss, A. On the motion of bubbles in capillary tubes. *J. Fluid Mech.* **1986**, *172*, 259–275.

122. Reinelt, D.; Saffman, P. The penetration of a finger into a viscous fluid in a channel and tube. *SIAM J. Sci. Stat. Comput.* **1985**, *6*, 542–561.

123. Hazel, A.L.; Heil, M. The steady propagation of a semi-infinite bubble into a tube of elliptical or rectangular cross-section. *J. Fluid Mech.* **2002**, *470*, 91–114.

124. Sarrazin, F.; Bonometti, T.; Prat, L.; Gourdon, C.; Magnaudet, J. Hydrodynamic structures of droplets engineered in rectangular micro-channels. *Microfluid. Nanofluid.* **2008**, *5*, 131–137.

125. Jousse, F.; Lian, G.; Janes, R.; Melrose, J. Compact model for multi-phase liquid–liquid flows in micro-fluidic devices. *Lab Chip* **2005**, *5*, 646–656.

126. Wörner, M. Numerical modeling of multiphase flows in microfluidics and micro process engineering: A review of methods and applications. *Microfluid. Nanofluid.* **2012**, *12*, 841–886.

127. Mashayek, F.; Pandya, R. Analytical description of particle/droplet-laden turbulent flows. *Prog. Energy Combust. Sci.* **2003**, *29*, 329–378.

128. Schmitt, M.; Stark, H. Marangoni flow at droplet interfaces: Three-dimensional solution and applications. *Phys. Fluids* **2016**, *28*, 012106, doi:10.1063/1.4939212.

129. Nguyen, N.T.; Lassemono, S.; Chollet, F.A. Optical detection for droplet size control in microfluidic droplet-based analysis systems. *Sens. Actuators B Chem.* **2006**, *117*, 431–436.

130. Lu, H.; Caen, O.; Vrignon, J.; Zonta, E.; El Harrak, Z.; Nizard, P.; Baret, J.C.; Taly, V. High throughput single cell counting in droplet-based microfluidics. *Sci. Rep.* **2017**, *7*, 1366, doi:10.1038/s41598-017-01454-4.

131. Chiou, P.Y.; Moon, H.; Toshiyoshi, H.; Kim, C.J.; Wu, M.C. Light actuation of liquid by optoelectrowetting. *Sens. Actuators A phys.* **2003**, *104*, 222–228.

132. Chiou, P.; Park, S.Y.; Wu, M.C. Continuous optoelectrowetting for picoliter droplet manipulation. *Appl. Phys. Lett.* **2008**, *93*, 221110.

133. Chiou, P.Y.; Chang, Z.; Wu, M.C. Droplet manipulation with light on optoelectrowetting device. *J. Microelectromech. Syst.* **2008**, *17*, 133–138.

134. Chuang, H.S.; Kumar, A.; Wereley, S.T. Open optoelectrowetting droplet actuation. *Appl. Phys. Lett.* **2008**, *93*, 064104, doi:10.1063/1.2970047.

135. Pei, S.N.; Valley, J.K.; Neale, S.L.; Jamshidi, A.; Hsu, H.Y.; Wu, M.C. Light-actuated digital microfluidics for large-scale, parallel manipulation of arbitrarily sized droplets. In Proceedings of the 2010 IEEE 23rd International Conference on IEEE Micro Electro Mechanical Systems (MEMS), Wanchai, Hong Kong, 24–28 Janruary 2010; pp. 252–255.

136. Pei, S.N.; Valley, J.K.; Neale, S.L.; Hsu, H.Y.; Jamshidi, A.; Wu, M.C. Rapid droplet mixing using light-actuated digital microfluidics. In Proceedings of the 2010 Conference on IEEE Lasers and Electro-Optics (CLEO) and Quantum Electronics and Laser Science Conference (QELS), San Jose, CA, USA, 16–21 May 2010; pp. 1–2.

137. Pei, S.N.; Wu, M.C. On-chip blade for accurate splitting of droplets in light-acuated digital microfluidics. In Proceedings of the 16th International Conference on Miniaturized Systems for Chemistry and Life Sciences, Okinawa, Japan, 28 October–1 November 2012.

138. Shekar, V.; Campbell, M.; Akella, S. Towards automated optoelectrowetting on dielectric devices for multi-axis droplet manipulation. In Proceedings of the 2013 IEEE International Conference on IEEE Robotics and Automation (ICRA), Karlsruhe, Germany, 6–10 May 2013; pp. 1439–1445.

139. Pei, S.N.; Valley, J.K.; Wang, Y.L.; Wu, M.C. Distributed circuit model for multi-color light-actuated opto-electrowetting microfluidic device. *J. Lightwave Technol.* **2015**, *33*, 3486–3493.

140. Tabeling, P. *Introduction to Microfluidics*; Oxford University Press: Oxford, UK, 2006.

141. Berthier, J. *Micro-Drops and Digital Microfluidics*; William Andrew: Norwich, NY, USA, 2012.

142. Ghenuche, P.; de Torres, J.; Ferrand, P.; Wenger, J. Multi-focus parallel detection of fluorescent molecules at picomolar concentration with photonic nanojets arrays. *Appl. Phys. Lett.* **2014**, *105*, 131102, doi:10.1063/1.4896852.

micromachines

MDPI

Review

Passive Mixing inside Microdroplets

Chengmin Chen [1], Yingjie Zhao [2], Jianmei Wang [1], Pingan Zhu [1], Ye Tian [3], Min Xu [1], Liqiu Wang [1,3] and Xiaowen Huang [1,*]

[1]   Energy Research Institute, Qilu University of Technology (Shandong Academy of Sciences), Jinan 250014, China; chencm@sderi.cn (C.C.); wangjm@sderi.cn (J.W.); zhupa@sderi.cn (P.Z.); xumin@sderi.cn (M.X.); lqwang@hku.hk (L.W.)
[2]   Key Laboratory of Biobased Polymer Materials, Shandong Provincial Education Department, School of Polymer Science and Engineering, Qingdao University of Science and Technology, Qingdao 266042, China; yz@qust.edu.cn
[3]   Department of Mechanical Engineering, The University of Hong Kong, Hong Kong, China; tianye@hku.hk
*   Correspondence: huangxiaowen2013@gmail.com; Tel.: +86-0531-88728328

Received: 31 January 2018; Accepted: 23 March 2018; Published: 1 April 2018

Abstract: Droplet-based micromixers are essential units in many microfluidic devices for widespread applications, such as diagnostics and synthesis. The mixers can be either passive or active. When compared to active methods, the passive mixer is widely used because it does not require extra energy input apart from the pump drive. In recent years, several passive droplet-based mixers were developed, where mixing was characterized by both experiments and simulation. A unified physical understanding of both experimental processes and simulation models is beneficial for effectively developing new and efficient mixing techniques. This review covers the state-of-the-art passive droplet-based micromixers in microfluidics, which mainly focuses on three aspects: (1) Mixing parameters and analysis method; (2) Typical mixing element designs and the mixing characters in experiments; and, (3) Comprehensive introduction of numerical models used in microfluidic flow and diffusion.

Keywords: passive mixing; microdroplets; multiphase; simulation model

1. Introduction

Microfluidics, working as a versatile, miniaturized, and integrated technology, has been widely used in many fields, such as chemical analysis, biological detection, drug screening, artificial photosynthesis, and microelectronics [1–5]. Mixing is necessary for these systems, particularly for those that involve heat transfer, mass transfer and chemical reaction. However, limited by the dynamic characters in microfluidics systems, most of the mixing efficiency is low and results in several problems. First, low mixing efficiency does not match the high-throughput requirements in quick analysis process, such as polymerase chain reaction (PCR) [6]. Second, low mixing efficiency may lead to heterogeneous mixture, causing low detection accuracy or poor product quality. Thus, the rapid homogeneous mixing is crucial in Lab-on-a-chip (LOC) platforms for widespread reactions covering biochemistry analysis, drug delivery, sequencing or synthesis of nucleic acids, cell activation, enzyme reactions, and protein folding.

According to the fluidic states in the mixers, mixing can be classified as single-phase mixing and droplet-based mixing [7]. In the single-phase mixing process, two or more reagents were injected into the microfluidic channels, and the mixing occurred by the diffusion between the fluid interfaces. Thus, the efficiency in single-phase mixing is limited by the diffusion flux, and dispersion of solutes along the channel is large (Figure 1a). Many methods have been introduced to overcome this limitation, such as via Split-and-Recombine [8], Chaotic flow [9], and nozzles [10]. When compared with single phase mixing, droplet-based mixing was put forward to overcome the drawbacks of low mixing efficiency in single-phase and high solutes dispersion because of internal recirculation and isolated environments [11,12]. In the droplets-based mixers, two or more kinds of reagents are driven into the channel independently and meet in the junction. The two immiscible phases used for the droplet generation are referred to as the continuous phase (medium in which droplets are generated) and dispersed phase (the droplet phase). With the cooperation of the geometry of the junction, the flow rates and the physical properties of the fluids the local flow field is determined, which leads to the interface deformation and droplet formation [13]. After the droplet has formed, the mixing begins. Because of its special performance when compared with the single phase mixing, droplet-based mixing plays a worthy role in the processes of chemical reactions [14–16], biological synthesis, and diagnostics [17–20], especially for miniaturized reactions or diagnostics [21].

However, it is also difficult to achieve a good mixing performance for droplets-based mixing in a straight channel, where the flow is laminar flow ($Re = 0.01~100$) [22]. The principle of intensifying the mixing is the increase of the diffusive flux between different disperse phase reagents, which is affected by the diffusion coefficient, interfacial surface area, and concentration gradient [23]. There are several devices designed to enhance the mixing in droplets, and they can be classified into two categories: active mixers and passive mixers. (1) In the active mixers, some external physical field is introduced to improve the efficiency via the external energy-induced eddy diffusion and bulk diffusion in the droplets [24]. Mixing in this mode relies on the materials' (those inside the fluid) response to the external physical field. Yesiloz et al. presented a microwave-based mixer, which heats the droplet by microwaves and induces non-uniform Marangoni stresses. In this mixer, highly viscous fluid (75% (w/w) glycerol solution) was selected as dispersed phase. By seeding with a fluorescent in half of the droplets, the mixing performance was investigated and the results showed that the mixing index reached as high as 97% within milliseconds [25] (Figure 2a). Bansal, et al. did some researches on the droplet mixing, depending on the non-axisymmetric oscillation patterns induced by actuation parameters in AC electrowetting, and the results show that the best mixing time in this system was approximately 2% of the diffusive mixing time [26] (Figure 2b). Besides these external energy sources, others are also used, such as magnetic field [27,28] or acoustics [29]. (2) In passive mixers, the mixing is achieved by the droplet movement only, and none of these structures employ external energy apart from the pump drive [30,31]. A simple scenario is making the droplets large enough to overfill the channel and exhibit a pancake shape, thus the mixing is intensified due to the shear forces between the wall and droplet interface. However, in this condition, the recirculation is enhanced only at one disperse phase side, and the mixing between two disperse phase reagents is limited. Therefore, some researchers introduced "Chaotic flow" into the droplet mixing process to break the limitation of the laminar regime. The most commonly used scheme is the channel deformation to realize the "Baker's transformation" in the droplets (Figure 1b–d) [32]. Figure 1b is the mixing process in one of curved channels. In these channel, the droplet stretched and folded inside the turns (sketched in Figure 1c), leading to asymmetrical recirculation in the droplets (shown in Figure 1d). In the deformation channels, the "Chaotic flow" in droplets is enhanced by stretching, folding, and rotation.

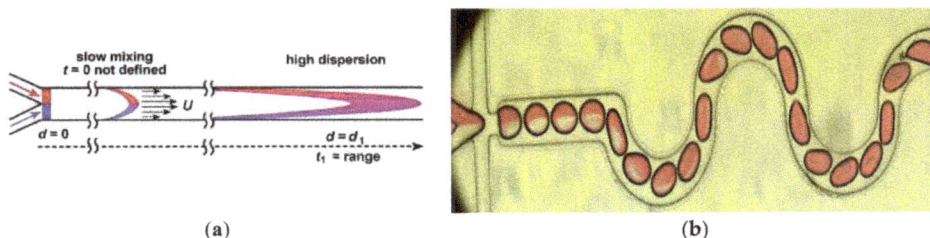

(a)   (b)

Figure 1. *Cont.*

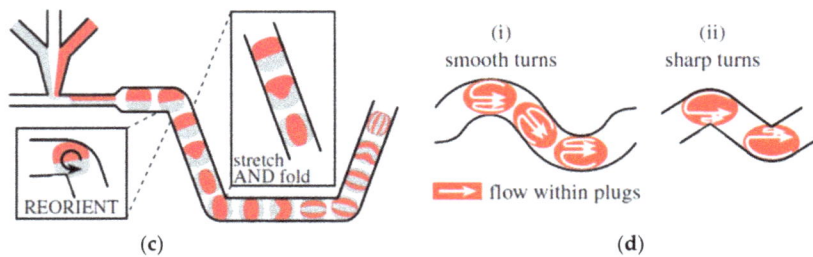

Figure 1. (**a**) Dispersion of solutes along the channel, reprinted with permission from [7]. (**b**) Microscopic image of the droplet in curved channel. (**c**,**d**) Recirculating flow scheme of Baker's transformation in curved channel, reprinted with permission from [32].

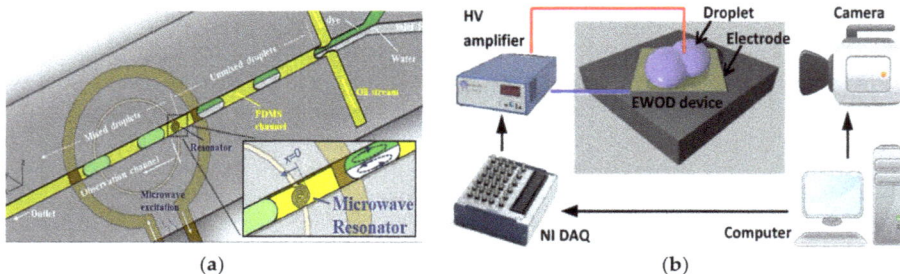

Figure 2. Droplet mixing in a microfluidic device (**a**) Mixing enhanced by adding a microwave heater: the heat induces non-uniform Marangoni stresses, leading to fast mixing, reprinted with permission from [25]. (**b**) Droplet mixing by alternating current (AC) electrowetting: the best mixing time was about 2% of the diffusive mixing time, reprinted with permission from [26].

Active mixers and passive mixers have their own advantages, respectively. For instance, active mixer presents their advantages in precise control and high efficiency, as well as high viscosity liquid mixing both in enclosed channel and on the substrate surface. They are also applicable for highly viscous fluid. Passive mixers show their advantages in terms of (1) reliable manipulation: the passive mixing relies on droplet movement in the immobile channel without any external energy and the external energy-induced instability [23]; (2) moderate reaction condition: external energy (e.g., heat or electric field) in the droplets may destroy some fragile molecules or deactivate some sensitive biomolecules; and, (3) easy fabrication of devices. With these merits, passive mixers are already applied in DNA hybridization analysis, polymerase chain reaction (PCR), cell activation, and chemical analysis [12,16,21] A brief comparison is showed in Table 1.

Table 1. Comparison of active and passive Mixers.

Categories	Principle	Mixing Time	Devices Fabrication	Stability of Operation	Application Scope
Active mixers	Disturbance caused by the external energy	Milliseconds	Complex, energy input including flow driven energy and mixing energy	Lower	The flow of response material or with response material
Passive mixers	Droplet movement in the immobile channel	Tens of milliseconds	Simple, only needs flow driven energy	Higher	All flow

In the following sections, we will briefly review the microdroplet-based passive mixer, including three parts: (1) Mixing parameters and analysis method, (2) Typical designs and mixing characters in experiments, and (3) Comprehensive introduction of numerical models that are used in the microfluidic flow and diffusion.

2. Characterization of Mixing in Microdroplets

Two parameters are important to characterize the mixing performance: the mixing time/mixing length and the distribution uniformity. When compared with the mixing time/mixing length, the distribution uniformity is more difficult to evaluate.

The most common and simplest way to measure the mixing uniformity is visible imaging, via flow visualization experiments with the aid of photometric, fluorescence intensity measurements [33,34], particle image velocimetry (PIV) measurements [35–37], high-speed color imaging [26], laser-induced fluorescence (LIF) [38,39], spontaneous Raman scattering (SRS), and many others [30]. However, the reliability of techniques and instruments limit the accuracy of visualization. Theoretical evaluation is a good solution for this problem. Most of evaluation methods are based on the principle of the non-homogeneous level of the tracer concentration distribution in a certain droplet. Standard deviation is an important parameter reflecting the uniformity of the tracer distribution, and sometimes it is adopted directly as a criterion of the mixing efficiency (Equation (1)) [40]. In Equation (1), σ_c is called standard deviation. When N is replaced by $N - 1$ [41,42], σ_s is called sample standard deviation, which is closer to the real deviation value in analyzing samples. The C is Concentration value, C_d is the reference concentration value.

$$\sigma_c = \sqrt{\frac{1}{N} \sum_{i=1}^{N}(C - C_d)^2} \text{ or } \sigma_s = \sqrt{\frac{1}{N-1} \sum_{i=1}^{N-1}(C - C_d)^2} \tag{1}$$

However, it is hard to compare different conditions using the standard deviation. During the last few years, several studies have been conducted on different micromixers that are aiming to characterize mixer performance. The widely used definition is the Danckwerts' segregation intensity index based on the variance, referred to as mixing index or mixing efficiency (m), as expressed by Equation (2) [43].

$$m = 1 - \frac{\sigma_c}{\sigma_{c,max}} \tag{2}$$

The concentration profiles are discrete functions. Therefore, Equation (2) is replaced by Equation (3) in experimental measurements [44,45],

$$m = 1 - \frac{\sqrt{\frac{1}{N} \sum_{i=1}^{N}(C - C_d)^2}}{C_d} \tag{3}$$

Other dimensionless numbers are also used based on the variance or standard deviation, such as intensity of segregation (IOS) [46] values and χ [34,47]. The symbol definition refers to the symbol list.

$$IOS = \frac{\sigma_{C\prime}^2}{C_d'(1 - C_d')} \tag{4}$$

$$\chi = \frac{\sigma_{c,t} - \sigma_{ref}}{\sigma_{c,0} - \sigma_{ref}} \tag{5}$$

In addition to these quantification methods, there are other novel characterization techniques. In the reaction systems, the reaction processes is accompanied by the enthalpy and kinetics change [48], so the dissipated energy is used to estimate the intensive mixing time. If the reaction is included, other indexes, such as pH value, are also used to evaluate the mixing performance according to the reaction type [49–51].

In simulations, mass fraction is usually used to estimate the mixing uniformity, since very detailed information can be obtained [44,52]. Similar to the experimental method, the same evaluation method is used. Because the mass fraction is a relative value in a certain area, the integral form is used in the equations (Equations (6) and (7)) [53,54]. In addition to the mass fraction, the particle tracer method that is based on the particle distribution in the droplet is usually used [55,56]. In this method, the particle location is varied with the mixing progress, and the uniformity of the particle location may represent the mixing uniformity to some extent.

$$m = \left(1 - \sqrt{\frac{\iint \left(C_f - C_{f,d}\right)^2 dA}{A \cdot C_d \left(C_{f,max} - C_{f,d}\right)}}\right) \cdot 100\% \tag{6}$$

$$m = \left(1 - \frac{\iint \left|C_f - C_{f,d}\right| dx dy}{\iint \left|C_{f,0} - C_{f,d}\right| dx dy}\right) \cdot 100\% \tag{7}$$

3. Micro-Mixers Design and Experiment Study

Passive mixing, without extra energy, relies on the molecular diffusion and chaotic advection in the droplet [57]. In the straight channels with two symmetrical streams, the formed droplets contain two symmetrical circulating flows on each half of a droplet, so it is difficult to mix together. Varying the channel geometries is the most effective strategy to enhance the mixing [57], and two methods were therefore proposed. One is to change the way of droplet formation, and the other is to change the way of droplet movement along the channel [12].

3.1. Mixing during Droplet Formation

To improve the mixing efficient, one can take full advantage of the internal recirculation inside the droplet during its generation.

For droplet mixing in the microfluidic devices, at least two dispersed phase inlets and one continuous phase inlet are required. The spatial location of the dispersed-phase inlets affects the fluid distribution inside the droplets, which may have a strong influence on the mixing performance [58]. The droplets that are formed by a bilateral symmetric fluid distribution structure take a long time to mix due to the low speed of diffusion between the two phases (Figure 3a), and some modification was proposed to improve the mixing in the cross junction (Figure 3b,c). Although the dispersion phase distribution is symmetrical in the droplets formed by these type, the droplets are more sensitive and the symmetry is more vulnerable to breaking, which leads to high mixing efficiency. Lin Bai studied the mixing efficiency of "Y-junction" in the mixing of ionic liquid (IL) droplets, and it is about 0.75 when compared with 0–0.4 in cross junction [12]. Wang, et al. studied the cross-shaped microchannel in Figure 3c. It is found that the IOS of the droplets generated in this channel has a low initial value [57]. Asymmetric dispersed phase inlets distribution [46,59,60], for example, "T-junction" (Figure 3d,e), is one of the possible methods to break the bilateral symmetric fluid distribution. Lin Bai also compared the mixing efficiency of "T-junction" with that of the cross section in the same conditions. It is verified that mixing efficiency is about 0.4–0.7, which is higher than 0–0.4 in cross junction. However, in this type, the reagent contacts the wall in the droplet formation process, which may have negative effects on the samples [54], so another asymmetric type was designed to overcome this drawback (Figure 3f,g). In these structures (Figure 3f,g), the advantage in enhancing the mixing performance is that the asymmetric design caused the vortex flows in the droplet formation, which leads to 1.5 times higher mixing index when compared with conventional flow-focusing structures [54].

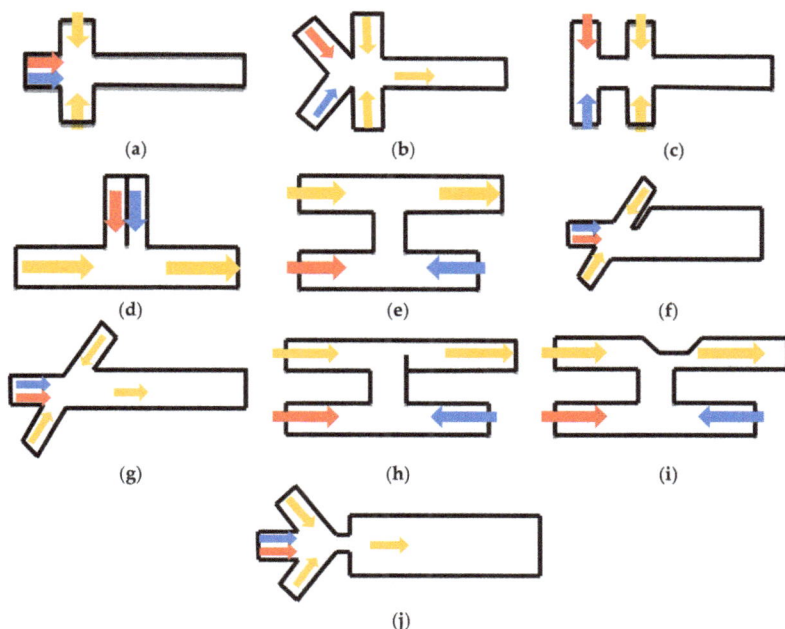

Figure 3. The schematic figures of the different generation sections. (**a–c**), three types of the asymmetric inlets model. (**d–i**), six types of the "convergent–divergent" model [61].

Another improved structure in the droplet formation section is to design a sudden shrink and enlarge the channel near the droplets formation location (Figure 3h–j). In "convergent–divergent" channels, the local flow speed is increased and the formation of the swirling structure in droplets is also sped up [12,54,62].

3.2. Mixing during Droplet Transportation

In the mixing process, fluid parameters and the microchannel structures impact the mixing performance obviously. Song, et al. [32] proposed a scaling law for the dependence of the mixing time $t \sim (aw/V)\ln(Pe)$ by the experiment of droplets mixing in meandering channels, where a is the dimensionless length of the plug measured relative to w. The discussion was useful to choose the proper operation condition. In a certain microchannel structure, a is important to the mixing performance. A smaller size ($a < 1$) results in a shorter mixing time/distance because of the high circulation speed. For the droplet that is large enough to contact with the channel walls, the mixing relies on the recirculation that is caused by the wall-induced shear stress. Wang, et al. [53] shows that when $1 < a < 2$, the asymmetric circulations make the disperse phase easy to mix in droplets moving in meandering channels. However, when $a > 2$, the asymmetric circulations had little effect on enhancing the mixing efficiency, which corresponds to the experimental results from Harshe, et al. [45]. In the following section, we focus on the mixing in the droplets with a diameter that is comparable to the channel width.

Microchannel structures deformation was widely used to break the symmetric recirculation in the droplets moving in the straight channel. The common methods are meandering channels (Figure 4a–f) or obstacle arrangements (Figure 4g) to break the symmetrical recirculation as well as to increase the effect of chaotic advection for mixing [34,53,56,58,63–65].

The dispersed phase reagent is reorientated within each turn of the curved channel. With the help of the reorientation, the mixing performance is enhanced. The obstacles function in the same way, which causes the asymmetrical circulation in the droplets. In some structures, the curved channel is combined with the "convergent–divergent" channel to enhance the mixing, and the results are good (Figure 4c,d). Tung et al. [66], did some experimental research about the droplets mixing in these structures in Figure 4c–e with the help of the high-tempo micro-particle image velocimetry (l-PIV). The results show that the mixing index increased eight times when compared with the straight microchannel at the same Reynolds number (Re = 2). The angle design of the turn is also very important for the enhancement of mixing efficiency (Figure 4e,f). Sarrazin, et al. [34] estimated the mixing performance in droplets using χ (Equation 5), and the results show that the channels with angles of 45° and 90° show good mixing efficiency (mixed within 10 ms) when compared with the straight channel or with an angle of 135° (mixing time is about 70 ms). Jiang, et al. [43] studied mixing efficiency in droplets moving in the channel with an angle of 60°, and the mixing time is about 18 ms when efficiency reached 80%. Besides meandering channels, other deforming channels were also presented (Figure 4g,h). In Figure 4g, the baffles were installed in the channel to change the position of the two independent circulation areas inside the large droplets [56], with the modification, the mixing efficiency decreased little with the droplet size increased. In Figure 4h, the droplet mixing is enhanced with the complex vortices generated by the droplet deformation when it crosses the compressed and expand areas alternatively [12].

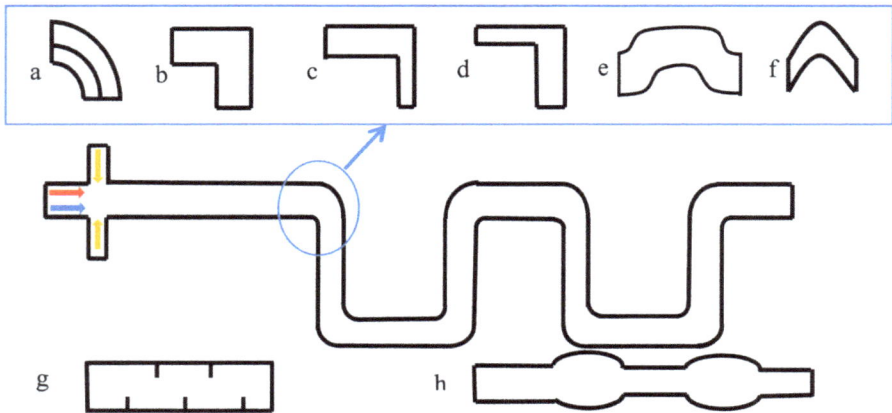

Figure 4. Deforming microchannels, (**a**–**f**) different structures of the meandering microchannel, (**g**) obstacle arrangements inside a microchannel, (**h**) the structure of a deforming channel.

Bai, et al. [12] gave detailed cooperation about the mixing performance inside IL droplets moving in various microchannels under the same flow conditions (Figure 5). Such a comparison demonstrates that although mixing inside IL (with viscosity of 66.4 mPa·s (25 °C)) is much more difficult than that in regular fluids, the design of combining the Y inlet (Figure 3b) and deforming channel (Figure 4h) could enhance the IL droplet mixing efficiency to high values that is close or even better than regular fluid droplets in common channels.

The most important principle to enhance the mixing performance is to break the symmetrical distribution of the disperse phase in the droplets. "Y" type and "T" type are the most used inlet structures with good mixing efficiency. In the meandering channels, small turn angles have good mixing performance. Good mixing performance can be achieved by combining the improved inlet structure and channel design.

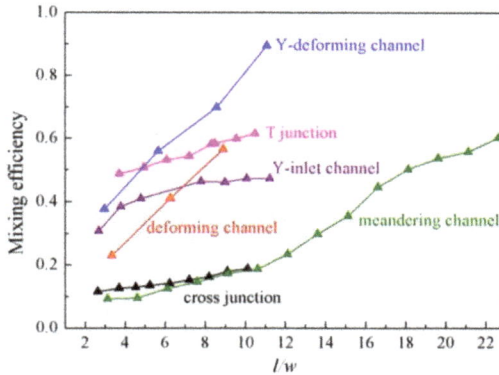

Figure 5. Mixing efficiency comparison in different micro-channels, reprinted with permission from [67].

4. Numerical Simulation

Computational fluid dynamics (CFD) is widely used to investigate the transport process in droplet-based mixing [68] for a comprehensive understanding of this process. In the droplet mixing processes, the multiphase model and the species transport model are used for investigation.

Droplet formation has been studied by a few researchers using the conventional CFD methods [68], including Volume of fluid method (VOF), Level Set method (LSM), Lattice Boltzmann method (LBM), and so on [54], while the species transport model is commonly used in the diffusion process. The most used multiphase flow models and species transport models are listed in Table 2.

Table 2. The difference of simulation models.

Method	Equations	Note	References
VOF and its improved methods	$\nabla \cdot V = 0$ $\frac{\partial \rho V}{\partial t} + \nabla(\rho V \cdot V) = -\nabla P + \rho g + \nabla \cdot \mu\left(\nabla V + \nabla V^T\right) + F$ $\frac{\partial \alpha}{\partial t} + \nabla(\alpha V) = 0$ $\alpha \begin{cases} 0, \, empty \, of \, A \, phase \\ 0 \sim 1, \, contain \, the \, interface \\ 1, \, filled \, of \, A \, phase \end{cases}$	Interface representation	[7,60,69–73]
LSE	$\nabla \cdot V = 0$ $\frac{\partial \rho V}{\partial t} + \nabla(\rho V \cdot V) = -\nabla P + \rho g + \nabla \cdot \mu\left(\nabla V + \nabla V^T\right) + F$ $\frac{\partial \varphi}{\partial t} + V \nabla \varphi = 0$ $\varphi(X,t) \begin{cases} d, \, if \, x \, in \, the \, liquid \, A \, phase \\ 0, \, if \, x \, in \, the \, interface \\ -d, \, if \, x \, in \, the \, liquid \, B \, phase \end{cases}$	Interface representation Interface $\varphi = 0$	[55,63]
LBM	$f_i^k(x + e_i\delta t, t + \delta t) = f_i^k(x,t) + \Omega_i^k + G_i^k$		[74,75]
Specie transport	$\frac{\partial C_i}{\partial t} + V \cdot \nabla C_i = D \nabla^2 C_i$		[76]

4.1. Volume-of-Fluid (VOF) Model

VOF [77] model is based on the fact that different phases are not interpenetrating, meaning that the fraction of the fluid volumes are addable in computational cells. It solves the surface flow with minimum consumption of computational resource, because this model solves a single set of Navier–Stokes differential equations for all of the phases and relies on the reconstruction of the interface by solving an advection equation (Table 1) [78]. However, because the VOF model suffers from the spurious velocities that are introduced from the computation of the mean curvature (Figure 6), its accuracy is poor in tracking the information of the interface [79]. Couplings with the improved interface tracking algorithm improve the VOF model accuracy [67,80].

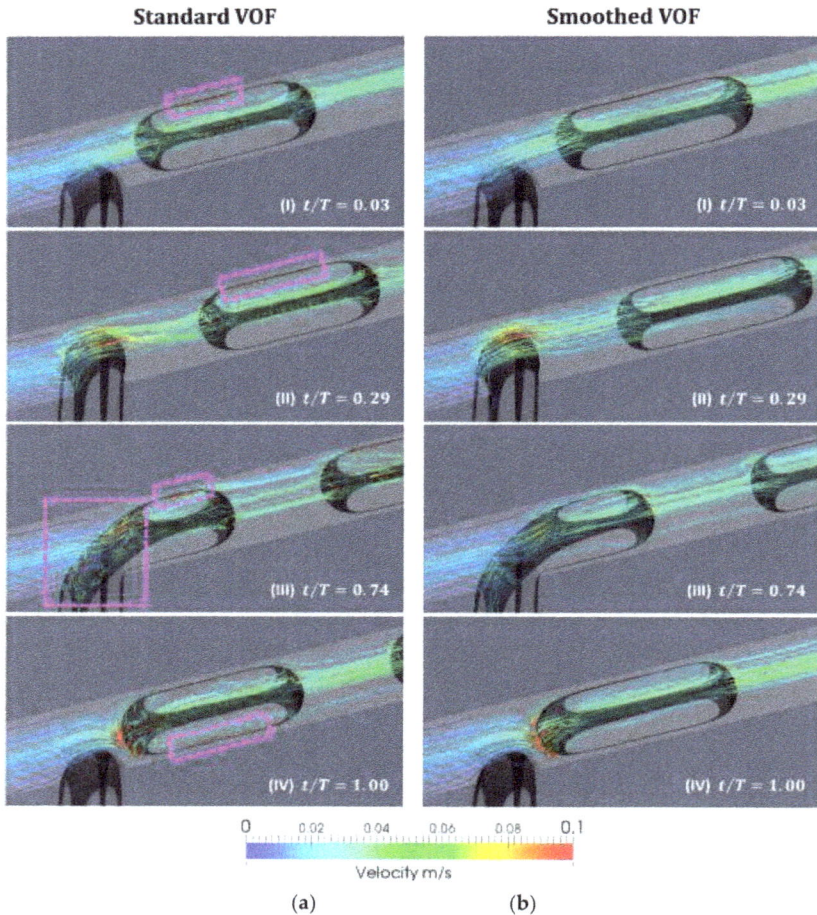

Figure 6. The flow fields near the interface during the droplet formation process, by (**a**) the Volume of fluid method (VOF) method and (**b**) the improved VOF method. Spurious velocities appeared near the interface in the VOF model. Reprinted from [67].

4.2. Level Set Method (LSM)

Level set method is an iterative, numerical technique to capture the interfaces and shapes within a fixed grid system [81]. In the level set method, an interface is represented by a contour of a smooth scalar field where $\phi = 0$, $\phi > 0$ and $\phi < 0$ represent two different phases.

When compared with the VOF model, the advantage of the level set method is the complex topological changes and computing with surface tension [82]. However, the level set function may be distorted by the flow field after some iteration steps, which leads to inaccurately approximated values on the interface (Figure 7). As a result, the simulation results may go against the mass conservation. To overcome this difficulty, some improved methods of the level set method is developed, such as dual-resolution LSM, sharp-interface LSM and conservative LSM [82,83].

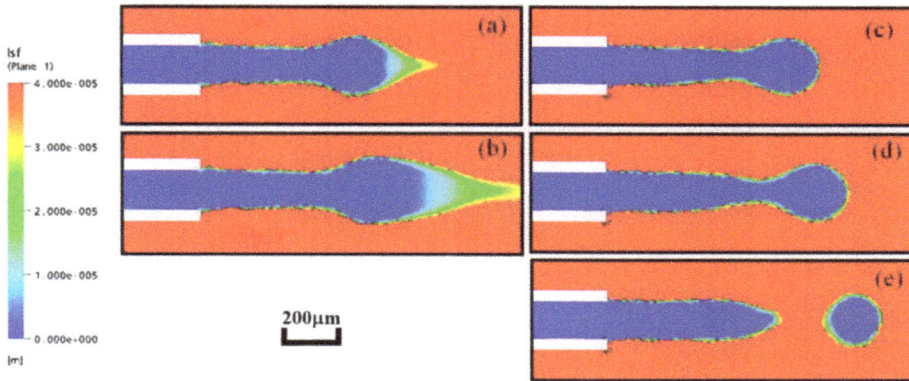

Figure 7. Simulation results comparison between the improved level set method and original level set method: (**a,b**), simulated by original level set method, (**c–e**), simulated by improved level set method. Reprinted with permission from [82].

4.3. VOF Coupled with Level Set Method (CLSVOF)

CLSVOF is a hybrid method combining both the level set method and the volume-of-fluid model [84]. This cooperation gets an improvement of the mass conservation as well as a more accurate interface-tracking (Figure 8). However, the hybrid method has a risk in numerical instability at the interface region when the interfacial tension is a dominant factor in complex geometries.

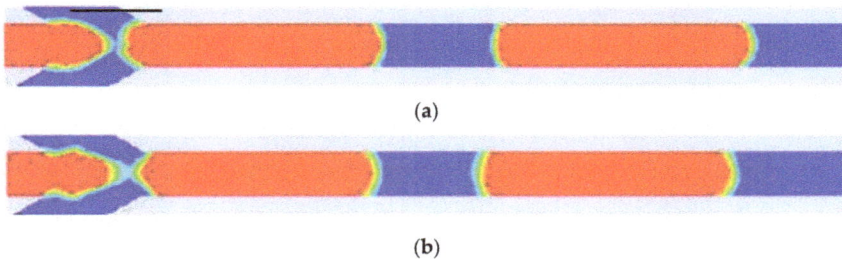

Figure 8. Comparison between (**a**) VOF Coupled with Level Set Method (CLSVOF) method (**b**) and VOF method. The interface in (**a**) is smoother than that in (**b**). Reprinted with permission from [73].

4.4. Lattice Boltzmann Method (LBM)

LBM, based on the Lattice Boltzmann equation, is developed from the discretized fluid model Lattice Gas Automata (LGA) [56,85,86]. The nature of the LB model is based on the assumption that molecular clusters are restricted at the discrete set of lattices and the molecular clusters act at each lattice side in two steps: collision and streaming [87]. LBM is positioned between the continuum level (described by Navier–Stokes equations) and the microscopic (molecular) level. The LBM could be classified to several types [88], such as color-fluid model [87,89], the pseudo-potential model [90], and the free-energy model [91]. LBM can successfully capture the motion and deformation of the interface and has been applied to simulate two-phase flows in microchannels. But, there are some drawbacks in these methods, such as the complexity and time-consumption in the coloring and recoloring step; the pseudo-potential model works well for low-density ratios only; and, the free-energy approach has a lack of Galilean invariance. That is why there are still other newly developed methods that are used in the droplet formation and mixing processes, according to different cases [92,93].

With the improvement of the simulation methods, the simulation results become more accurate, which plays an important role in the study of droplet mixing. The previous work has shown that, when compared with experiment results, each method can capture the droplets formation dynamics and the internal velocity fields inside the droplets successfully. When combined with species transport method, the simulation error in mixing efficiency is also acceptable. Yang, et al. compared the experiment and simulation results with LSM method, and in same condition, when mixing efficiency is 0.9. The mixing time of experiment and simulation is about 0.025 s and 0.027 s, respectively [63]. Jiang, et al. also did the comparison using LSM mothed, and the result is that when the mixing efficiency is 0.8, the mixing times of experiment and simulation are about 0.014 s and 0.018 s, respectively [43]. Wang, et al. studied the IOS of droplets mixing in meandering channels, the simulation results have good concordance with experimental results in the range of $0 < l/w < 10$, even in some place, the IOS value coincided very well [58].

When we do the simulation works, we should choose the best models according to the simulation conditions, such as the accuracy of results, the flow situation, and the purpose. With the help of the simulation, the modifications of the structures and parameter studies are time- and cost-saving. Table 3 lists some numerical studies on mixing in recent years.

Table 3. Some studies on simulation of mixing performance in recent years.

Mixing Mechanism	Conditions	Evaluate Method	Mixing Performance	Model	Reference
Baffle in channel	$Pe \sim 102$; $Ca = 0.0008 \sim 0.08$; $Re = 0.2 \sim 20$	Standard deviation	When $l/w = 6$, σ_c is about 0	LBM	[56]
Asymmetric inlets	$Re = 0.39 \sim 2.93$	Mixing index	When $t = 14$ ms, 0.90	VOF	[59]
	$Ca = 0.06 \sim 0.006$; $Re = 0.1 \sim 0.001$	Mixing index	When Ca = 0.06, 0.90	COMSOL Multiphysics	[54]
Serpentine microchannel	$V = 1.11$ m^3/s; 2.22 m^3/s; $D = 100$ μm/120 μm	Particle trajectories/time	0.08s	LSM	[55]
	$V(A) = 0.005 \sim 0.04$ m/s; $V(B) = 0.01$ m/s	Mixing index	$L/w = 16$, 0.90	LSM	[63]
	$D = 50$ mm; $Re = 3.10$; $Ca = 0.0036$	Mixing index	When $L/w = 16 \sim 32$, 0.90	VOF	[53]
Converging shape	$Ca \sim 0.02$; $V(A) = 100$ μL·min^{-1}; $V(B) = 10$ μL·min^{-1}	IOS	When $L/w = 10$ 0.2~0.4	LBM	[46]
	$Ca \sim 0.022$; $Re \sim 2.5$	IOS	When $L/w = 10$, 0.5	LBM	[58]

Note: A: continuous phase; B: dispersed phase.

5. Conclusions

When compared with the active mixers, passive mixers have the advantages of lower cost, simple device design, and reduced power input, which make the passive mixers widely applicable. In this paper, we gave a brief review of the experimental and simulation results of the passive mixers based on droplets. Also, we systematically analyzed the quantification methods, newly developed types of mixers and the simulation methods. With the improved experimental devices and simulation methods, more detailed mixing principles and structure designs have been present. The results presented in this review will shed light on further study of droplet based-passive mixers. In the future, the innovative mixing technology will be more widely used integrated with other steps from sample-in to result-out. Besides that, there is still need for some studies on mixing for special applications, such as high viscosity, high heat release/absorb from reaction in the droplet, and so on.

Acknowledgments: This work was supported by the Shandong Province Natural Science Foundation (No. ZR2017LEE018) and Shandong Academy of Sciences Foundation (2013QN016 and 2017BSHZ007).

Author Contributions: Chengmin Chen and Xiaowen Huang wrote the paper; Yingjie Zhao and Jianmei Wang organized figures; Ye Tian, Pingan Zhu, Min Xu, Liqiu Wang contributed to the paper structure and sentences revise.

Conflicts of Interest: The authors declare no conflict of interest.

Abbreviation

Symbol

m	Mixing index
N	The number of the samples
C	Concentration value
A	Area
l	Length of channel
w	width of channel
D	Channel diameters
σ_c	Standard deviation of concentration
σ_s	Sample standard deviation
C_d	The reference concentration value
t	Time, s
C'	Normalized concentration
C'_d	The statistical average value of normalized concentration in the entire droplet
C_f	Mass fraction
$C_{f,d}$	The reference Mass fraction
ρ	Density
μ	Viscosity
g	Gravitational constant
V	Velocity
$\mathbf{V}$	Vector of velocity
Pe	Peclet number
C_a	Capillary number
e	Vector of lattice direction
Re	Reynolds number
F	Interaction force
P	pressure
Ω_i^k	Collision operator
G_i^k	External force term
α	Volume fraction in VOF model

Subscrips

0	The initial condition
max	The maximum value
t	The condition of certain time
∞	The condition in well mixing section

References

1. Whitesides, G.M. Overview the origins and the future of microfluidics. *Nature* **2006**, *442*, 368–373. [CrossRef] [PubMed]

2. Huang, X.; Wang, J.; Li, T.; Wang, J.; Xu, M.; Yu, W.; El Abed, A.; Zhang, X. Review on optofluidic microreactors for artificial photosynthesis. *Beilstein J. Nanotechnol.* **2018**, *9*, 30–41. [CrossRef] [PubMed]

3. Erickson, D.; Sinton, D.; Psaltis, D. Optofluidics for energy applications. *Nat. Photonics* **2011**, *5*, 583–590. [CrossRef]

4. Huang, X.; Hui, W.; Hao, C.; Yue, W.; Yang, M.; Cui, Y.; Wang, Z. On-site formation of emulsions by controlled air plugs. *Small* **2014**, *10*, 758–765. [CrossRef] [PubMed]
5. Huang, X.; Zhu, Y.; Zhang, X.; Bao, Z.; Lei, D.Y.; Yu, W.; Dai, J.; Wang, Y. Clam-inspired nanoparticle immobilization method using adhesive tape as microchip substrate. *Sens. Actuators B Chem.* **2016**, *222*, 106–111. [CrossRef]
6. Hayes, C.J.; Daltona, T.M. Microfluidic droplet-based PCR instrumentation for high-throughput gene expression profiling and biomarker discovery. *Biomol. Detect. Quantif.* **2015**, *4*, 22–32. [CrossRef] [PubMed]
7. Madadelahi, M.; Shamloo, A. Droplet-based flows in serpentine microchannels: Chemical reactions and secondary flows. *Int. J. Multiph. Flow* **2017**, *97*, 186–196. [CrossRef]
8. Hossain, S.; Kim, K.Y. Mixing analysis in a three dimensional serpentine split-and-recombine micromixer. *Chem. Eng. Res. Des.* **2015**, *100*, 95–103. [CrossRef]
9. Chen, C.Y.; Lin, C.Y.; Hu, Y.T.; Cheng, L.Y.; Hsu, C.C. Efficient micromixing through artificial cilia actuation with fish-schooling configuration. *Chem. Eng. J.* **2015**, *259*, 391–396. [CrossRef]
10. Sayah, A.; Gijs, M.A.M. Simulation and fabrication of a three-dimensional microfluidic mixer in a monolithic glass substrate. *Procedia Eng.* **2015**, *120*, 229–232. [CrossRef]
11. Song, H.; Tice, J.D.; Ismagilov, R.F. A microfluidic system for controlling reaction networks in time. *Angew. Chem. Int. Ed.* **2003**, *42*, 767–772. [CrossRef] [PubMed]
12. Bai, L.; Fu, Y.; Yao, M.; Cheng, Y. Enhancement of mixing inside ionic liquid droplets through various micro-channels design. *Chem. Eng. J.* **2018**, *332*, 537–547. [CrossRef]
13. Baroud, C.N.; Gallaire, F.; Dangla, R. Dynamics of microfluidic droplets. *Lab Chip* **2010**, *10*, 2032–2045. [CrossRef] [PubMed]
14. Yang, C.H.; Lin, Y.S.; Shih, M.C.; Chiu, H.C.; Huang, K.S. Droplet-based microfluidic technology applications in polymer science. *Curr. Proteom.* **2014**, *11*, 92–97. [CrossRef]
15. Jerzy, B.J.; Kotowicz, M. Application of new chemical test reactions to study mass transfer from shrinking droplets and micromixing in the rotor-stator mixer. *Chem. Process Eng.* **2017**, *38*, 477–489.
16. Dressler, O.; Solvas, X.C.I.; deMello, A.J. Chemical and biological dynamics using droplet-based microfluidics. *Annu. Rev. Anal. Chem.* **2017**, *10*, 1–24. [CrossRef] [PubMed]
17. Ishida, T.; McLaughlin, D.; Tanaka, Y.; Omata, T. First-come-first-store microfluidic device of droplets using hydrophobic passive microvalves. *Sens. Actuators B Chem.* **2018**, *254*, 1005–1010. [CrossRef]
18. Juárez, J.; Brizuela, C.A.; Martínez-Pérez, I.M. An evolutionary multi-objective optimization algorithm for the routing of droplets in digital microfluidic biochips. *Inf. Sci.* **2018**, *429*, 130–146. [CrossRef]
19. Bera, N.; Bhattacharya, B.B.; Majumder, S. Simulation-based method for optimum microfluidic sample dilution using weighted mix-split of droplets. *IET Comput. Dig. Tech.* **2016**, *10*, 119–127. [CrossRef]
20. Yan, Y.; Guo, D.; Luo, J.; Wen, S. Numerical simulation of droplet dynamic behaviors in a convergent microchannel. *BioChip J.* **2013**, *7*, 325–334. [CrossRef]
21. Cao, J.; Köhler, J.M. Droplet-based microfluidics for microtoxicological studies. *Eng. Life Sci.* **2015**, *15*, 306–317. [CrossRef]
22. Zhu, P.; Tang, X.; Tian, Y.; Wang, L. Pinch-off of microfluidic droplets with oscillatory velocity of inner phase flow. *Sci. Rep.* **2016**, *6*, 31436. [CrossRef] [PubMed]
23. Hessel, V.; Löwe, H.; Schönfeld, F. Micromixers—A review on passive and active mixing principles. *Chem. Eng. Sci.* **2005**, *60*, 2479–2501. [CrossRef]
24. Lu, L.H.; Ryu, K.S.; Liu, C. A magnetic microstirrer and array for microfluidic mixing. *J. Microelectromech. Syst.* **2002**, *11*, 462–469.
25. Yesiloz, G.; Boybay, M.S.; Ren, C.L. Effective thermo-capillary mixing in droplet microfluidics integrated with a microwave heater. *Anal. Chem.* **2017**, *89*, 1978–1984. [CrossRef] [PubMed]
26. Shubhi, B.; Prosenjit, S. Mixing enhancement by degenerate modes in electrically actuated sessile droplets. *Sens. Actuators B Chem.* **2016**, *232*, 318–326.
27. Teste, B.; Jamond, J.; Ferraro, D.; Viovy, J.L.; Malaquin, L. Selective handling of droplets in a microfluidic device using magnetic rails. *Microfluid. Nanofluid.* **2015**, *19*, 141–153. [CrossRef]
28. van Reenen, A.; de Jong, A.M.; den Toonder, J M.J.; Prins, M.W.J. Integrated lab-on-chip biosensing systems based on magnetic particle actuation–a comprehensive review. *Lab Chip* **2014**, *14*, 1966–1986. [CrossRef] [PubMed]

29. Yeo, L.Y.; Friend, J.R. Ultrafast microfluidics using surface acoustic waves. *Biomicrofluidics* **2009**, *3*, 120002. [CrossRef] [PubMed]

30. Zeng, Y.; Jiang, L.; Zheng, W.; Li, D.; Yao, S.; Yao, S.; Qu, J.Y. Quantitative imaging of mixing dynamics in microfluidic droplets using two-photon fluorescence lifetime imaging. *Opt. Lett.* **2011**, *36*, 2236–2238. [CrossRef] [PubMed]

31. Zhu, P.; Wang, L. Passive and active droplet generation with microfluidics: A review. *Lab Chip* **2016**, *17*, 34–75. [CrossRef] [PubMed]

32. Song, H.; Bringer, M.R.; Tice, J.D.; Gerdts, C.J.; Ismagilov, R.F. Experimental test of scaling of mixing by chaotic advection in droplets moving through microfluidic channels. *Appl. Phys. Lett.* **2003**, *83*, 4664–4666.

33. Dong, Z.; Zhao, S.; Zhang, Y.; Yao, C.; Yuan, Q.; Chen, G. Mixing and residence time distribution in ultrasonic microreactors. *AIChE J.* **2017**, *63*, 1404–1418. [CrossRef]

34. Sarrazin, F.; Prat, L.; Di Miceli, N.; Cristobal, G.; Link, D.R.; Weitz, D.A. Mixing characterization inside microdroplets engineered on a microcoalescer. *Chem. Eng. Sci.* **2007**, *62*, 1042–1048. [CrossRef]

35. Saroj, S.K.; Asfer, M.; Sunderka, A.; Panigrahi, P.K. Two-fluid mixing inside a sessile micro droplet using magnetic beads actuation. *Sens. Actuators A Phys.* **2016**, *244*, 112–120. [CrossRef]

36. Dore, V.; Tsaoulidis, D.; Angeli, P. Mixing patterns in water plugs during water/ionic liquid segmented flow in microchannels. *Chem. Eng. Sci.* **2012**, *80*, 334–341. [CrossRef]

37. Carrier, O.; Ergin, F.G.; Li, H.-Z.; Watz, B.B.; Funfschilling, D. Time-resolved mixing and flow-field measurements during droplet formation in a flow-focusing junction. *J. Micromech. Microeng.* **2015**, *25*, 081014. [CrossRef]

38. Carroll, B.; Hidrovo, C. Experimental investigation of inertial mixing in colliding droplets. *Heat Transf. Eng.* **2013**, *34*, 120–130. [CrossRef]

39. Yeh, S.-I.; Fang, W.F.; Sheen, H.J.; Yang, J.T. Droplets coalescence and mixing with identical and distinct surface tension on a wettability gradient surface. *Microfluid. Nanofluid.* **2013**, *14*, 785–795. [CrossRef]

40. Davanlou, A.; Kumar, R. Passive mixing enhancement of microliter droplets in a thermocapillary environment. *Microfluid. Nanofluid.* **2015**, *19*, 1507–1513. [CrossRef]

41. Shamloo, A.; Madadelahi, M.; Akbari, A. Numerical simulation of centrifugal serpentine micromixers and analyzing mixing quality parameters. *Chem. Eng. Process. Process Intensif.* **2016**, *104*, 243–252. [CrossRef]

42. Shamloo, A.; Vatankhah, P.; Akbari, A. Analyzing mixing quality in a curved centrifugal micromixer through numerical simulation. *Chem. Eng. Process. Process Intensif.* **2017**, *116*, 9–16. [CrossRef]

43. Jiang, L.; Zeng, Y.; Zhou, H.; Qu, J.Y.; Yao, S. Visualizing millisecond chaotic mixing dynamics in microdroplets: A direct comparison of experiment and simulation. *Biomicrofluidics* **2012**, *6*, 012810. [CrossRef] [PubMed]

44. Cortes-Quiroz, C.A.; Azarbadegan, A.; Zangeneh, M. Effect of channel aspect ratio of 3-d t-mixer on flow patterns and convective mixing for a wide range of reynolds number. *Sens. Actuators B Chem.* **2017**, *239*, 1153–1176. [CrossRef]

45. Harshe, Y.M.; van Eijk, M.J.; Kleijn, C.R.; Kreutzer, M.T.; Boukany, P.E. Scaling of mixing time for droplets of different sizes traveling through a serpentine microchannel. *RSC Adv.* **2016**, *6*, 98812–98815. [CrossRef]

46. Zhao, S.; Wang, W.; Zhang, M.; Shao, T.; Jin, Y.; Cheng, Y. Three-dimensional simulation of mixing performance inside droplets in micro-channels by lattice boltzmann method. *Chem. Eng. J.* **2012**, *207–208*, 267–277. [CrossRef]

47. Zivkovic, V.; Ridge, N.; Biggs, M.J. Experimental study of efficient mixing in a micro-fluidized bed. *Appl. Therm. Eng.* **2017**, *127*, 1642–1649. [CrossRef]

48. Romano, M.; Pradere, C.; Sarrazin, F.; Toutain, J.; Batsale, J.C. Enthalpy, kinetics and mixing characterization in droplet-flow millifluidic device by infrared thermography. *Chem. Eng. J.* **2015**, *273*, 325–332. [CrossRef]

49. Guo, X.; Fan, Y.; Luo, L. Mixing performance assessment of a multi-channel mini heat exchanger reactor with arborescent distributor and collector. *Chem. Eng. J.* **2013**, *227*, 116–127. [CrossRef]

50. Lin, X.; Zhang, J.; Wang, K.; Luo, G. Determination of the micromixing scale in a microdevice by numerical simulation and experiments. *Chem. Eng. Technol.* **2016**, *39*, 909–917. [CrossRef]

51. Bai, L.; Zhao, S.; Fu, Y.; Cheng, Y. Experimental study of mass transfer in water/ionic liquid microdroplet systems using micro-lif technique. *Chem. Eng. J.* **2016**, *298*, 281–290. [CrossRef]

52. Cortes-Quiroz, C.A.; Azarbadegan, A.; Zangeneh, M. Evaluation of flow characteristics that give higher mixing performance in the 3-d t-mixer versus the typical t-mixer. *Sens. Actuators B Chem.* **2014**, *202*, 1209–1219. [CrossRef]

53. Wang, J.; Wang, J.; Feng, L.; Lin, T. Fluid mixing in droplet-based microfluidics with a serpentine microchannel. *RSC Adv.* **2015**, *5*, 104138–104144. [CrossRef]

54. Filatov, N.A.; Belousov, K.I.; Bukatin, A.S.; Kukhtevich, I.V.; Evstrapov, A.A. The study of mixing of reagents within a droplet in various designs of microfluidic chip. *J. Phys. Conf. Ser.* **2016**, *741*, 012052. [CrossRef]

55. Özkan, A.; Erdem, E.Y. Numerical analysis of mixing performance in sinusoidal microchannels based on particle motion in droplets. *Microfluid. Nanofluid.* **2015**, *19*, 1101–1108. [CrossRef]

56. Zhao, S.; Riaud, A.; Luo, G.; Jin, Y.; Cheng, Y. Simulation of liquid mixing inside micro-droplets by a lattice boltzmann method. *Chem. Eng. Sci.* **2015**, *131*, 118–128. [CrossRef]

57. Capretto, L.; Cheng, W.; Hill, M.; Zhang, X. Micromixing within microfluidic devices. *Top. Curr. Chem.* **2011**, *304*, 27–68. [PubMed]

58. Wang, W.; Shao, T.; Zhao, S.; Jin, Y.; Cheng, Y. Experimental and numerical study of mixing behavior inside droplets in microchannels. *AIChE J.* **2012**, *59*, 1801–1813. [CrossRef]

59. Li, Y.; Reddy, R.K.; Kumar, C.S.; Nandakumar, K. Computational investigations of the mixing performance inside liquid slugs generated by a microfluidic t-junction. *Biomicrofluidics* **2014**, *8*, 054125. [CrossRef] [PubMed]

60. Chandorkar, A.; Palit, H. Simulation of droplet dynamics and mixing in microfluidic devices using a VOF-based method. *Sens. Transducers J.* **2009**, *7*, 136–149.

61. Vitae, A.A.; Kim, K.Y. Convergent–divergent micromixer coupled with pulsatile flow. *Sens. Actuat. B* **2015**, *211*, 198–205.

62. Ahn, S.; Kim, D.W.; Kim, Y.W.; Yoo, J.Y. Generation of Janus droplets for enhanced mixing in microfluidics. *Int. J. Precis. Eng. Manuf.* **2010**, *11*, 799–802. [CrossRef]

63. Yang, L.; Li, S.; Liua, J.; Cheng, J. Fluid mixing in droplet-based microfluidics with t junction and convergent-divergent sinusoidal microchannels. *Electrophoresis* **2017**, *39*, 512–520. [CrossRef] [PubMed]

64. Mazutis, L.; Baret, J.C.; Griffiths, A.D. A fast and efficient microfluidic system for highly selective one-to-one droplet fusion. *Lab Chip* **2009**, *9*, 2665–2672. [CrossRef] [PubMed]

65. Song, H.; Ismagilov, R.F. Millisecond kinetics on a microfluidic chip using nanoliters of reagents. *J. Am. Chem. Soc.* **2003**, *125*, 14613–14619. [CrossRef] [PubMed]

66. Tung, K.Y.; Li, C.C.; Yang, J.T. Mixing and hydrodynamic analysis of a droplet in a planar serpentine micromixer. *Microfluid. Nanofluid.* **2009**, *7*, 545–557. [CrossRef]

67. Soh, G.Y.; Yeoh, G.H.; Timchenko, V. Improved volume-of-fluid (VOF) model for predictions of velocity fields and droplet lengths in microchannels. *Flow Meas. Instrum.* **2016**, *51*, 105–115. [CrossRef]

68. Wörner, M. Numerical modeling of multiphase flows in microfluidics and micro process engineering: A review of methods and applications. *Microfluid. Nanofluid.* **2012**, *12*, 841–886. [CrossRef]

69. Hoang, V.T.; Lim, J.; Byon, C.; Park, J.M. Three-dimensional simulation of droplet dynamics in planar contraction microchannel. *Chem. Eng. Sci.* **2018**, *176*, 59–65. [CrossRef]

70. Li, X.B.; Li, F.C.; Yang, J.C.; Kinoshita, H.; Oishi, M.; Oshima, M. Study on the mechanism of droplet formation in t-junction microchannel. *Chem. Eng. Sci.* **2012**, *69*, 340–351. [CrossRef]

71. Soh, G.Y.; Yeoh, G.H.; Timchenko, V. Numerical investigation on the velocity fields during droplet formation in a microfluidic t-junction. *Chem. Eng. Sci.* **2016**, *139*, 99–108. [CrossRef]

72. Saha, A.A.; Mitra, S.K. Effect of dynamic contact angle in a volume of fluid (VOF) model for a microfluidic capillary flow. *J. Colloid Interface Sci.* **2009**, *339*, 461–480. [CrossRef] [PubMed]

73. Dang, M.; Yue, J.; Chen, G. Numerical simulation of Taylor bubble formation in a microchannel with a converging shape mixing junction. *Chem. Eng. J.* **2015**, *262*, 616–627. [CrossRef]

74. Fu, Y.; Bai, L.; Zhao, S.; Zhang, X.; Jin, Y.; Cheng, Y. Simulation of reactive mixing behaviors inside micro-droplets by a lattice Boltzmann method. *Chem. Eng. Sci* **2018**, *181*, 79–89. [CrossRef]

75. Riaud, A.; Zhao, S.; Wang, K.; Cheng, Y.; Luo, G. Lattice-boltzmann method for the simulation of multiphase mass transfer and reaction of dilute species. *Phys. Rev. E Stat. Nonlinear Soft Matter Phys.* **2014**, *89*, 053308. [CrossRef] [PubMed]

76. Mandal, M.M.; Aggarwal, P.; Nigam, K.D.P. Liquid–liquid mixing in coiled flow inverter. *Ind. Eng. Chem. Res.* **2011**, *50*, 13230–13235. [CrossRef]

77. Olsson, E.; Kreiss, G. A conservative level set method for two phase flow. *J. Comput. Phys.* **2005**, *210*, 225–246. [CrossRef]

78. Swapna, S.; Rabha, V.V.B. Volume-of-fluid(VOF) simulations of rise of single/multiple bubbles in sheared liquids. *Chem. Eng. Sci.* **2010**, *65*, 527–537.

79. Bashir, S.; Rees, J.M.; Zimmerman, W.B. Simulations of microfluidic droplet formation using the two-phase level set method. *Chem. Eng. Sci.* **2011**, *66*, 4733–4741. [CrossRef]

80. Pozzetti, G.; Peters, B. A multiscale DEM-VOF method for the simulation of three-phase flows. *Int. J. Multiph. Flow* **2018**, *99*, 186–204. [CrossRef]

81. Watanabe, Y.; Saruwatari, A.; Ingram, D.M. Free-surface flows under impacting droplets. *J. Comput. Phys.* **2008**, *227*, 2344–2365. [CrossRef]

82. Lan, W.; Li, S.; Wang, Y.; Luo, G. CFD simulation of droplet formation in microchannels by a modified level set method. *Ind. Eng. Chem. Res.* **2014**, *53*, 4913–4921. [CrossRef]

83. Sharma, A. Level set method for computational multi-fluid dynamics: A review on developments, applications and analysis. *Sadhana* **2015**, *40*, 627–652. [CrossRef]

84. Griebel, M.; Klitz, M. Clsvof as a fast and mass-conserving extension of the level-set method for the simulation of two-phase flow problems. *Numer. Heat Trans. Part B Fundam.* **2016**, *71*, 1–36. [CrossRef]

85. Fu, Y.; Bai, L.; Zhao, S.; Bi, K.; Jin, Y.; Cheng, Y. Droplet in droplet: LBM simulation of modulated liquid mixing. *Chem. Eng. Sci.* **2016**, *155*, 428–437. [CrossRef]

86. Liu, H.; Valocchi, A.J.; Kang, Q. Three-dimensional lattice Boltzmann model for immiscible two-phase flow simulations. *Phys. Rev. E Stat. Nonlinear Soft Matter Phys.* **2012**, *85*, 046309. [CrossRef] [PubMed]

87. Fu, Y.; Bai, L.; Bi, K.; Zhao, S.; Jin, Y.; Cheng, Y. Numerical study of Janus droplet formation in microchannels by a lattice boltzmann method. *Chem. Eng. Process. Process Intensif.* **2017**, *119*, 34–43. [CrossRef]

88. Halliday, I.; Hollis, A.P.; Care, C.M. Lattice Boltzmann algorithm for continuum multicomponent flow. *Phys. Rev. E Stat. Nonlinear Soft Matter Phys.* **2007**, *76*, 026708. [CrossRef] [PubMed]

89. Leclaire, S.; Reggio, M.; Trépanier, J.-Y. Progress and investigation on lattice Boltzmann modeling of multiple immiscible fluids or components with variable density and viscosity ratios. *J. Comput. Phys.* **2013**, *246*, 318–342. [CrossRef]

90. Liu, H.; Zhang, Y. Droplet formation in microfluidic cross-junctions. *Phys. Fluids* **2011**, *23*, 082101. [CrossRef]

91. van der Zwan, E.; van der Sman, R.; Schroen, K.; Boom, R. Lattice boltzmann simulations of droplet formation during microchannel emulsification. *J. Colloid Interface Sci.* **2009**, *335*, 112–122. [CrossRef]

92. Galusinski, C.; Vigneaux, P. On stability condition for bifluid flows with surface tension: Application to microfluidics. *J. Comput. Phys.* **2008**, *227*, 6140–6164. [CrossRef]

93. Shardt, O.; Derksen, J.J.; Mitra, S.K. Simulations of Janus droplets at equilibrium and in shear. *Phys. Fluids* **2014**, *26*, 012104. [CrossRef]

micromachines

MDPI

Review

Advances of Optofluidic Microcavities for Microlasers and Biosensors

Zhiqing Feng [1] and Lan Bai [2,*]

[1] College of Physics and Materials Engineering, Dalian Nationalities University, Dalian 116600, China; fzq@dlnu.edu.cn
[2] College of Mechanical and Electronic Engineering, Dalian Nationalities University, Dalian 116600, China
* Correspondence: bailan@dlnu.edu.cn

Received: 29 January 2018; Accepted: 6 March 2018; Published: 9 March 2018

Abstract: Optofluidic microcavities with high Q factor have made rapid progress in recent years by using various micro-structures. On one hand, they are applied to microfluidic lasers with low excitation thresholds. On the other hand, they inspire the innovation of new biosensing devices with excellent performance. In this article, the recent advances in the microlaser research and the biochemical sensing field will be reviewed. The former will be categorized based on the structures of optical resonant cavities such as the Fabry–Pérot cavity and whispering gallery mode, and the latter will be classified based on the working principles into active sensors and passive sensors. Moreover, the difficulty of single-chip integration and recent endeavors will be briefly discussed.

Keywords: microcavities; optofluidic dye lasers; whispering gallery modes; biosensors

1. Introduction

In recent years, optofluidic microcavities have been developed, becoming a key element of microfluidic platforms. Many kinds of microcavities with high Q value and small mode volume have been obtained by using microfabrication technology [1–9] thanks to their excellent light confinement for a long time in a small volume. They enjoy a significant enhancement of light-matter interaction and narrow resonance linewidth, making them favorable for optofluidic microlasers and biochemical sensing applications.

Depending on the light confinement mechanism, the microcavities are generally divided into two categories: Fabry–Pérot (FP) cavities and whispering gallery mode (WGMs) cavities. The materials can be optical fibers, microcapillaries, polymers, silicon or glass substrates. As illuminated in Figure 1, there are usually several common FP microcavities, as defined by their geometric shapes: plane-plane mirror type (PPFP), concave-concave mirror type (CCFP) and plane-concave mirror type (PCFP). Figure 2 shows more kinds of WGM microcavities, including microring, microdisk, microtoroid, microsphere, microbubble and microbottle.

Optofluidic dye lasers are formed by integrating microcavities and gain medium into proper microfluidic circuits or devices. Lots of gain media have been used, such as dyes, quantum dots, rare earth ions, labeled-DNA, fluorescent proteins, chlorophyll solutions, etc. As miniaturized light sources, the optofluidic dye lasers have the merits of low threshold and high Q factor. In addition, they have made significant advances in other aspects such as full bio-compatibility, mode selecting between single and multi-mode, lasing wavelength tunability, and so on. As biochemical sensing elements, the optofluidic lasers usually obtain much higher sensitivity than traditional detecting techniques [10–17]. For sensing applications, they are also described as active resonator sensors.

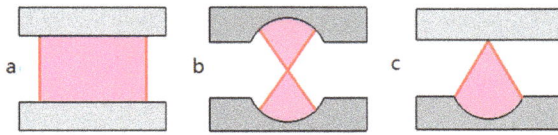

Figure 1. Three configurations of Fabry–Pérot (FP) microcavity, (**a**) plane-plane mirror type (PPFP). (**b**) concave-concave mirror type (CCFP). (**c**) plane-concave mirror type (PCFP).

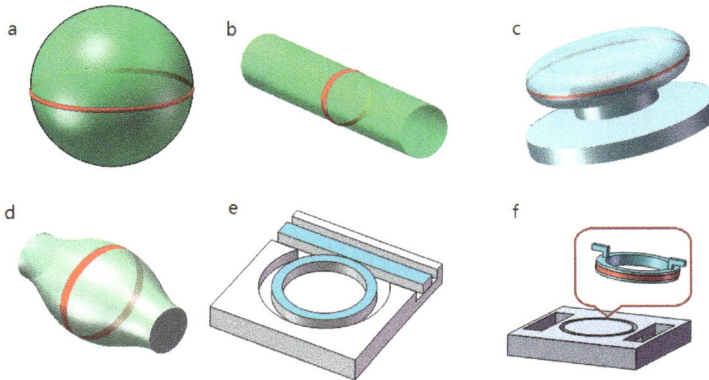

Figure 2. Classical configurations of whispering gallery mode (WGM)-based microcavities. (**a**) Microsphere; (**b**) cylindrical ring; (**c**) microdisk or microtoroid; (**d**) microbottle; (**e**) monolithic solid core microring; (**f**) monolithic liquid core microring.

Optofluidic microcavities can also be worked as passive resonator sensors without gain medium. Recent research has mainly focused on the WGM-based resonators [18–27]. The sensing ability of WGM resonators is characterized by the figure of merit using the Q-factor value to mode volume ratio (Q/V). The sensitivity can be improved by increasing the Q value or decreasing the resonators' mode volume. Through ultrafine laser processing, many microscale resonators have been fabricated, and their detection limits have been successfully reduced down to several kDa molecular weight for single particles [19–25].

This article is not the first to review the topic of optofluidic microcavities. In 2010, Y. Chen gave an in-depth review on the physical theory and the development status of optofluidic microcavities [28]. However, the research of optofluidic microcavities has made substantial progress since then, and many inspiring studies have emerged. It is now a good time to update the latest research progress of optofluidic microcavities in two application areas: optofluidic dye lasers and microcavity-based biosensors.

2. Optofluidic Microcavities for Dye Lasers

2.1. Fabry–Pérot Cavity Dye Lasers

The FP cavities are easy to fabricate. For example, optical fiber end faces or glass slides can be used to constitute the PPFP cavity; microscale concaves (or concave arrays) made by laser machining on planar substrates can form the PCFP cavity [29–35]. Figure 3 shows the PCFP cavity array structure made by Wang [29]. By depositing the Bragg reflection dielectric layers, the Q value was enhanced to 5.6×10^5. When the cavity length was 31 µm, the laser threshold was lowered to 0.09 µJ·mm^{-2}. When the cavity length was shortened to 8 µm, the excitation threshold was increased to 0.5 µJ·mm^{-2}, and single mode lasing was observed at 599 nm. The Lahoz group reported another simple design of a

PPFP dye laser [31] which could be excited by a low-power continuous-wave (CW) laser diode with the threshold of 1.3 μJ·mm^{-2}. As a sensor, it could be operated in laser mode or fluorescence mode by changing the excitation laser intensity.

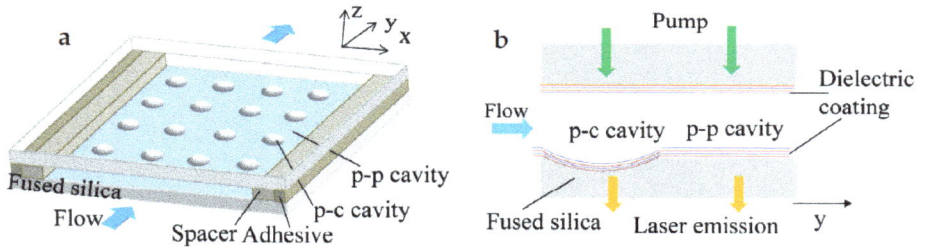

Figure 3. (**a**) Schematic of the optofluidic laser array based on the PCFP and PPFP cavities. (**b**) Details of the experimental setup employing both the PCFP and the PPFP cavity on the same fused silica chip [29].

Gerosa [33] constructed all-fiber high-repetition-rate microfluidic dye lasers by welding the optical fibers and the capillary tubes. The excitation threshold was about 1 μJ by using 532-nm, 300-ps 1-kHz pulse laser. The structure is illuminated in Figure 4. Some key features of the FP cavity dye lasers are listed in Table 1.

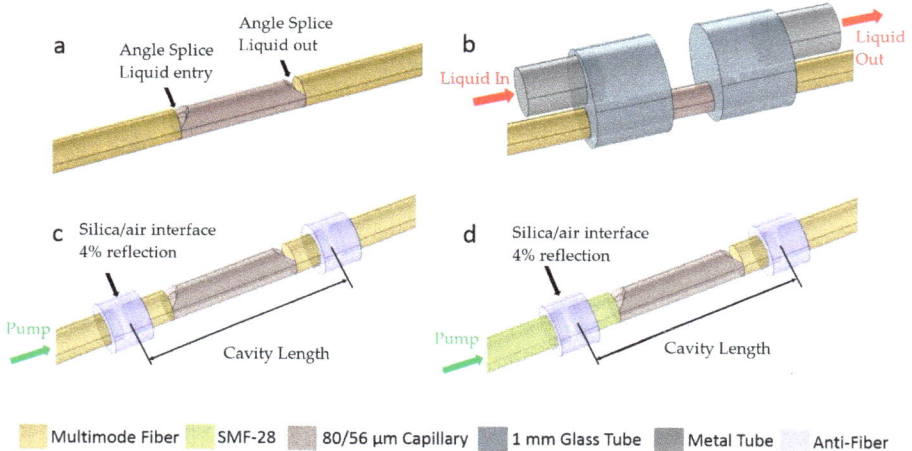

Figure 4. Device schemes of the all-fiber high-repetition-rate microfluidic dye lasers. (**a**) Angle-cleaved capillary spliced to conventional fibers, allowing for liquid flow; (**b**) a whole device, including the pressure cells formed by glass tubes and their connection to liquid reservoirs via metal tubes; (**c**) multimode laser cavity, including air gaps for the feedback via Fresnel reflection; (**d**) few-mode laser cavity with similar air gaps but with a small-core fiber (SMF-28) in one side to provide modal filtering. Anti-fiber is the capillary (inner diameter 128 μm) used to generate the air gaps [33].

Table 1. List of the optofluidic lasers based on the Fabry–Pérot microcavities.

Ref.	Cavity Configuration	Cavity Length (μm)	Q-Factor	Threshold (μJ·mm^{-2})	Lasing Mode	Gain Materials	Cavity Materials
[29]	PCFP	31	5.6×10^5	0.09	Mutlimode	R6G	Fused Silica substrate
		8	5.6×10^5	0.7	Single mode	R6G	
[30]	PCFP	39	4×10^5	0.13	Mutlimode	R6G	Fused Silica substrate
[31]	PPFP	150		9.6	Mutlimode	MB	Fused Silica plate
[32]	PPFP	165		1.3	Mutlimode	IgG-Atto488 complex	Fused Silica plate
[33]	PPFP	~10,000		1	Mutlimode	Rh640	Fiber, caplilary

2.2. WGM Dye Lasers

The WGM dye lasers are obtained by the combination of a liquid or solid state gain medium and WGM microresonators. If the medium around the microresonator has positive optical gain, the evanescent wave of the WGMs would interact with the medium to generate WGM laser emission. Various forms of resonators have been demonstrated, such as microring, microsphere, microbubble, microdisc, microtoroid, and microbottle [36–62]. Cylindrical and planar microring are more popular due to their simple configurations which confine the photon propagation in the quasi-two-dimensional space. Recently, a solid or hollow microbottle based on microcylinder or microcapillary structures has been proposed for the WGM lasers. The advantage of the microbottle structure is that it has multiple non-degenerated modes along the axis of the revolution, which are convenient for modes selection. Single-mode lasing can be realized by spatial pump engineering [42–44]. Here, we list in Table 2 the main features of the recently proposed WGM dye lasers.

Table 2. List of the optofluidic lasers based on the WGM microcavities.

Ref.	Cavity Configuration	Cavity Length (μm)	Q-Factor	Threshold	Lasing Mode	Gain Materials	Cavity Materials
[37]	Cylindrical ring resonator	~410	2.6×10^6	5.9 μJ/mm^2	Single mode, 386.75 nm	LD390	Microcapillary, glass solid cylinder
[38]	Cylindrical ring resonator	59.9–90.9		16–44 nJ/pulse	~10 nm tunable range, axial pumping	R6G, RhB	Hollow core microstructured fiber
[39]	Cylindrical ring resonator	17.4		664 nJ·mm^{-2}	Single longitudinal mode, lateral pumping	R6G	Hollow core microstructured fiber
[40]	Cylindrical ring resonator	157,393		Several tens μJ/mm^2	Mutlimode, 520–560 nm	Ribo-flavin	Microcapillary
[41]	Cylindrical ring resonator	157	6000	1.2 μJ	Mutlimode, 600–615 nm	Nile red dye	Microcapillary, polymer
[48]	Cylindrical ring resonator	393	~106	23 μJ/mm^2	Mutlimode, 510–520 nm	eGFP	Bare SM-28 fiber
[2]	Monolithic liquid-core ring resonator	534	3.3×104	15 μJ/mm^2	Mutlimode, 570–580 nm	R6G	Glass
[42]	Microbottle	9–19		10–20 μW/mm^2	Single mode, 580–620 nm, tunable	R6G	Microfiber, polymer
[4]	Microbottle	534		~3.6 mW	Multimode, 1530–1540 nm	Er: Yb doped glass	glass
[51]	Droplet	323	5800		Multimode, 590–610 nm	R6G	Dichloro-methane and epoxy resin

In the WGM lasers, the carrier utilizes the hollow microstructured fibers, the microcapillaries or the planar microrings on chips, and the gain medium liquid is filled in or flowed through. In some other designs, the dye-doped polymer is coated on the inner or outer wall of resonator to form the microring resonator dye lasers using the side or axial pumping. The cavity length of cylindrical resonators can be further reduced by tapering.

In general, the laser output of the microring resonator lasers is spatially divergent. By the WGM mode-coupling between the lasing resonator and another solid cylinder resonator, the emission direction could be limited to a certain range, thus forming the directional emission. As shown in Figure 5, Tu reported the uses of thin-walled capillary and solid cylinder to construct the coupled ring resonator dye laser [37]. Ultraviolet single-frequency laser emission was generated with a pump threshold of 5.9 μJ·mm^{-2}. The laser emission was mainly in two directions with a divergence of 10.5°. The single mode lasing of different power was realized by changing the position of the two resonators.

In another work, Lee [40] used riboflavin water solution as the gain medium to construct the microring optofluidic lasers based on microcapillary tube and optical fiber, respectively. Riboflavin has good biocompatibility as compared to other organic dyes. Multimode lasing at 520–560 nm band was obtained by side pumping of optical parametric oscillator (OPO) laser. The threshold was several tens to one hundred μJ·mm^{-2}.

In addition to the liquid gain media, the solid gain layer was also proposed by coating dye-doped polymer on the surface of microcapillary resonators and was first demonstrated by Francois [41]. Usually, the laser features vary with the thickness of gain layer and the solution refractive index in the capillary. The proper thickness range of polymer was 600–800 nm. The multimode lasing of 590–630 nm was generated under the excitation of 532 nm laser by the side pumping. The excitation threshold was lowered to 1.2 μJ (thickness of 800 nm) and 16 μJ (thickness of 600 nm), respectively.

Hollow-core micro-structured optical fibers have a smaller scale than the microcapillaries, and have thus lower internal connection losses. They are often used as miniaturized resonators by tapering. The cavity length is different along the axis. This feature could be applied to frequency tuning. Recently, Liu group [38] proposed a tunable microring dye laser, in which RhB and R6G were used as gain media. The threshold of 16–44 nJ/pulse was obtained by the axial pumping. The tuning range was 10 nm. Besides, Yu [39] constructed a single longitudinal mode optofluidic microring laser by the hollow microstructure fiber. The effective cavity length was about 109.3 μm. The dye fluid was injected into the hollow fiber. The threshold was lowered to 664 nJ·mm^{-2} by the side pumping. Different dyes were used for laser emission of different wavelengths.

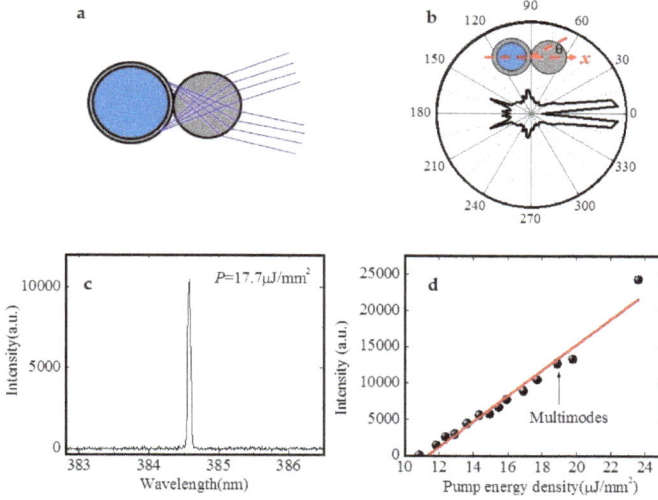

Figure 5. (**a**) Schematic of the focused light rays from the cylinder. (**b**) Far-field distribution of light. (**c**) Single wavelength lasing under lower pump power. (**d**) Plot of the relation between the output intensity and pump energy density [37].

In general, the planar liquid-core microring resonator requires a liquid-core waveguide channel to connect to the liquid microring resonator channel to transport the gain medium, which inevitable results in a decrease of the Q-factor. By using the three-dimensional (3D) direct-writing of the femtosecond laser, the gain medium inlet and outlet channels could be designed in the non-WGM area and the high Q-factor could be maintained using the 3D pipeline design. Monolithic microring laser on glass substrate was first reported by Fan's group [2]. The ring cavity had the inner radius, the outer radius and depths of 150 μm, 170 μm and 40 μm, respectively. As shown in Figure 6, R6G dye was dissolved in a quinoline solution with refractive index 1.62 to act as the gain medium. It was pumped by nanosecond pulses which were generated by a 532-nm optical parametric oscillator (OPO) laser. Since the fluid refractive index was bigger than the glass, the WGM wave was mainly confined in the fluid close to the outer edge. The lasing spectrum was multimode and the lasing threshold was approximated 15 μJ·mm^{-2}.

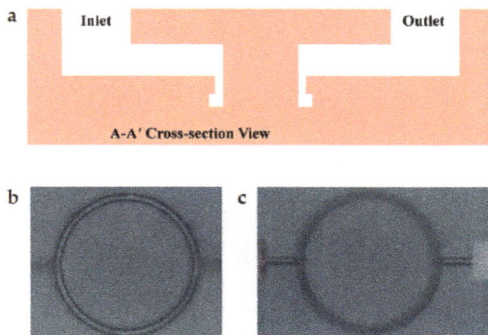

Figure 6. (**a**) Cross-sectional view of the monolithic optofluidic ring resonator; (**b**,**c**) Micrographs of the bottom and the top by focusing on the ring and the fluid delivery channel, respectively [2].

The WGM dye lasers based on the microbottle have multi-wavelengths distributed along the axis. A single WGM lasing mode could be obtained by the spatial modulation approach of pumping, which may result from the laser-interference excitation field. Gu's group [42] proposed WGM lasing in dye-doped polymer microbottle resonators, as shown in Figure 7. The pump energy distribution profile along the axis could be rearranged by adjusting the angle between the two excitation beams. The lasing might be single mode by tuning the space of the fringes along the axis and the frequency could be tuned by applying a tensile stress along the fiber axis.

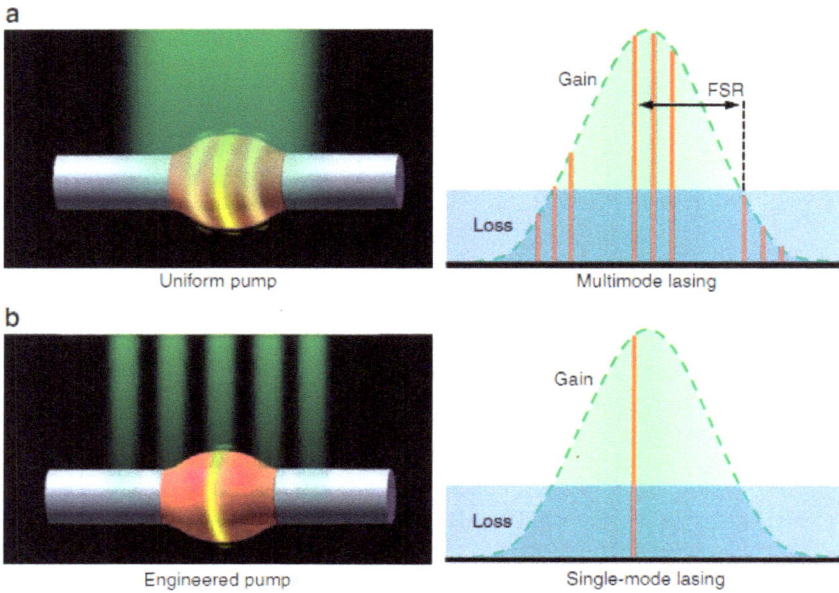

Figure 7. WGM dye laser based on the microbottle structure. (**a**) Uniform pump and multimode lasing; (**b**) modulation pumping and single mode lasing [42].

Optofluidic lasers with a single molecular layer of gain was first reported by Fan's group [48,49]. The gain layers used enhanced green fluorescent protein (eGFP), dye-labeled bovine serum albumin (BSA) and dye-labeled DNA, and were assembled on the surface of ring resonators by the surface immobilization biochemical methods. This is a very interesting work for high sensitivity surface bio-detection.

The immiscible dye droplets suspended in the solution are excellent disk-like optical resonators. They also produce lasing output under proper pumping and can be tuned by the solution interfacial tension. Yang [51] used the inkjet print technology to inject a gain medium solution to float on the water to form a fully liquid WGM microlaser. The tension was changed by the concentration of soap water.

3. Optofluidic Microcavities for Biosensors

The physical mechanism of the bio-sensing using the optical microcavities is that the electric field distribution is changed by the variation of refractive index of surrounding medium. The redistribution of electric field would alter the cavities' resonance mode, which in turn would vary the time (or frequency) domain features of signal light, such as a resonant peak shift, resonant mode splitting, broadening, and intensity variation, etc. By detecting these optical features, the concentration of species can be obtained, which is highly related to the refractive index of sample solution. Especially for the WGM cavities with high Q factor, standing waves are formed around the ring due to the long light travelling time (or distance). Even if a single bio-particle of nanoscale is locally attached to the cavities' surface, it would also rearrange the electric field distribution because of highly enhanced light-particle interaction. Recently, much research attention has been put on the WGM-based detection of single particles, such as virus, DNA and single proteins [19–25,63–79]. Some reviews had made detailed descriptions of the theories and the recent biosensing applications of optical microresonators [27,28,67,68]. Here, we will just review the bio-sensing from the two supplementary aspects: microcavity-based active biosensing and microcavity-based passive biosensing.

3.1. Microcavity-Based Active Biosensing

When the target concentration is extremely low, the traditional fluorescent intensity-based bio-detecting methods hardly work due to the low signal intensity and various types of noises. For example, enzyme-linked immunosorbent assay (ELISA) kits rely on the intensity of fluorescence generated from the product of the enzyme-substrate reaction so as to quantify the targets attached to the solid surface of ELISA kits. The detection limits are usually sub-$\mu g \cdot L^{-1}$ for most targets' bulk solutions and are hard to decrease further due ro the influences of nonspecific bindings, the auto-fluorescence of materials and leakage of excitation light. To improve the detection limit of ELISA, Fan's group [12] incorporated the PPFP cavities into the ELISA kits, in which fluorescence was confined and resonated to lasing by detecting the lasing onset time to obtain the concentration of interleukin-6 solutions. The threshold of the laser was below 320 $\mu J \cdot mm^{-2}$ using the 532-nm OPO pulsed laser pumping. The detection limit was reduced to 1 $fg \cdot mL^{-1}$ and the dynamic range was extended to 10^6. A similar detection method was applied to the photocatalytic reaction by the same group [13], which constructed an optofluidic catalytic laser for ultra-sensitive sulfide ion detection.

Besides this, the fluorescence resonance energy transfer (FRET) process can be incorporated into the microcavity to form a laser-based sensing platform, which would greatly improve the sensitivity of bio-sensing [15,16]. More descriptions of the theories of FRET laser-based sensing can be also found in [17].

Moreover, Ren [63,64] proposed an optofluidic laser for high-sensitivity and low-detection-limit sensing of refractive index, which obtained the sensitivity of 3874 nm/RIU and the noise equivalent detection limit of 2.6×10^{-6} RIU. Zhang [65] improved the refractive index sensitivity of the microring laser by two orders of magnitude via the strong coupling between the ring laser and the fluidic microtube.

3.2. Microcavity-Based Passive Biosensing

For the passive bio-sensors based on microcavities, the researchers mainly focused on WGM-based resonators due to the powerful detection abilities of surface bio-reactions. Different configurations of resonators have been implemented for label-free bio-sensing, such as cylindrical ring, bottle, bubble-like, disc or toroidal, and planar liquid core ring [19–26]. As the passive sensors, no gain medium is needed and thus no fluorescence or lasing is produced. An external light source (white source or tunable laser) is used to couple photons into the cavities by the taper fiber or the waveguide. By monitoring the shift or splitting the resonant peaks of transmission spectra, analyte concentration or molecule attaching can be detected [69–84]. As the optical confinement elements, the cavities with high Q factor would greatly enhance the light-matter interaction and would result in high sensitivity. However, there are some potential problems such as light source fluctuation, temperature variation, large background caused by the low couple efficiency of excitation, detector noises, etc. All of these factors would deteriorate the detection limit significantly.

Hybrid microcavities have been reported to utilize the plasmon resonances to further enhance the light–matter interaction [85–89]. Advanced signal processing techniques such as self-reference differential detection, frequency locking (or phase locking detection) are developed to improve the signal and noise ratio (SNR) [90–98]. The detection limit is reduced to the level of ~5 kDa for a single bio-particle. An excellent example of this kind of work was reported by Zhang [91], who developed a self-referenced differential mode sensing method. It used two resonant modes in the same microbottle resonator to reduce the measurement noises from the exciting source fluctuation. The detection limit of 10 fg·mL^{-1} for bovine serum albumin molecules was obtained. Su [94] used the laser frequency locking technique to improve the SNR of microtoroid resonators, obtaining the detection of single nanoparticles of 2.5 nm in radius and 15.5 kDa molecular weight. In addition to the detection of liquid concentration and single particles, these sensors could also be used for gas sensing applications [99,100].

Although optofluidic microcavities based on WGM have great potential in sensing, the integration on a single chip remains a big challenge. For example, high-quality 3D microdisk (or microtoroid) resonators could be easily fabricated by laser processing on glass or silicon substrates, but the photonic waveguides that are necessary for delivering the probe light are hard to fabricate. Taper fibers are usually applied and more extra processing is needed. Yet the uncertainty regarding the geometric parameters of taper fiber and the gap between them are severe hindrances to bulk production. Resonators made by microcapillaries or hollow fibers are also difficult to integrate on a monolithic chip owing to their large sizes and structural fragility. Recently, Schmidt's group [101] has developed an innovative approach called *lab-in-a-tube*, which integrates numerous rolled-up components into a single device on a chip. Figure 8 shows a TiO_2 microtubular optical resonator as a result of the rolling-up of a 2D planar membrane deposited on the substrates due to the surface stress. In addition, the resonator is integrated with vertical-sited SU-8 polymer waveguide. The geometric parameters of microtube, the waveguide and the gap between them were well defined and controllable at nanoscale. The test results showed that the resonators had good sensing performance and excellent optical coupling efficiency with an extinction ratio of 32 dB over the communication band. Other materials such as SiO_2 was also developed by the same group [102,103]. These studies made an important contribution to the research of optofluidic monolithic integration.

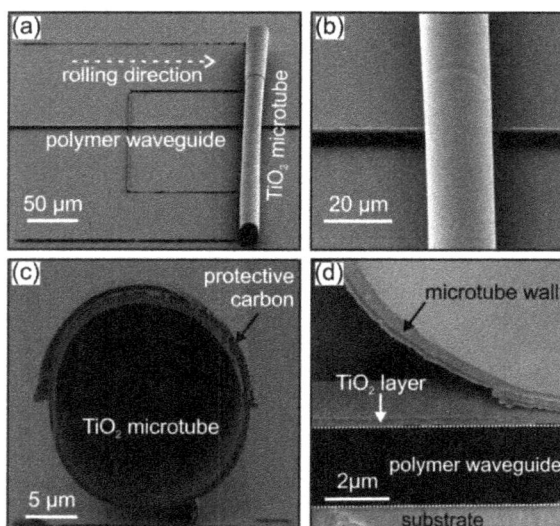

Figure 8. Lab-in-a-tube system made of rolled-up TiO_2 microresonators integrated with polymer waveguides. (**a**) Microtube was rolled-up by a U-shaped pattern. (**b**) Close-up view of the tube that is connected tightly with the polymer waveguide to ensure optimal optical coupling. (**c**) Compact winding layers of the tube wall were revealed by a FIB cut with the protection of a carbon layer. (**d**) The FIB cut image at the waveguide revealed the compact tube wall in the vicinity of the polymer waveguide [101].

4. Conclusions

Optofluidic microcavities have found wide applications and are still expanding their application areas rapidly. Here, we have summarized the recent progress in the areas of microlasers and biosensors. Generally, the optofluidic microlasers are developing toward high Q factors with a low threshold, small volume, easy mode controllability and wide tunability. In addition, the high performance of microresonators improves the light-matter interaction and thus greatly enhances the sensing abilities and the scope of applications. By means of process improvement, structure integration and detection method innovation, new microcavity devices with higher performance are presented continuously. It is expected that more practical devices will be developed for lasers, biosensors and other applications.

Acknowledgments: This project is supported by the Educational Committee Foundation of Liaoning Province (L2015126).

Author Contributions: Zhiqing Feng collected and analyzed the data, prepared the spreadsheets and figures and drafted the manuscript. Lan Bai conceived and revised the manuscript.

Conflicts of Interest: The authors declare no conflict of interest.

References

1. Wang, M.; Lin, J.T.; Xu, Y.X.; Fang, Z.-W.; Qiao, L.-L.; Liu, Z.-M.; Fang, W.; Cheng, Y. Fabrication of high-Q microresonators in dielectric materials sing a femtosecond laser: Principle and applications. *Opt. Commun.* **2017**, *395*, 249–260. [CrossRef]
2. Chandrahalim, H.; Chen, Q.; Said, A.A.; Dugan, M.; Fan, X. Monolithic optofluidic ring resonator lasers created by femtosecond laser nanofabrication. *Land Chip* **2015**, *15*, 2335–2340. [CrossRef] [PubMed]
3. Simoni, F.; Bonfadini, S.; Spegni, P.; Lo, T.S.; Lucchetta, D.E.; Criante, L. Low threshold Fabry-Pérot optofluidic resonator fabricated by femtosecond laser micromachining. *Opt. Express* **2016**, *24*, 17416–17423. [CrossRef] [PubMed]
4. Ward, J.M.; Yang, Y.; Chormaic, S.N. Glass-on-glass fabrication of bottle-shaped tunable microlasers and their applications. *Sci. Rep.* **2016**, *6*, 25152. [CrossRef] [PubMed]
5. Tada, K.; Cohoon, G.; Kieu, K.; Mansuripur, M.; Norwood, R.A. Fabrication of high-Q microresonators using femtosecond laser micromachining. In Proceedings of the Lasers and Electro-Optics (CLEO), San Jose, CA, USA, 6–11 May 2012; pp. 1–2.
6. Lin, J.; Xu, Y.; Tang, J.; Wang, N.; Song, J.; He, F.; Fang, W.; Cheng, Y. Fabrication of three-dimensional microdisk resonators in calcium fluoride by femtosecond laser micromachining. *Appl. Phys. A* **2014**, *116*, 2019–2023. [CrossRef]
7. Lin, J.; Xu, Y.; Fang, Z.; Wang, M.; Song, J.; Wang, N.; Qiao, L.; Fang, W.; Cheng, Y. Fabrication of high-Q lithium niobate microresonators using femtosecond laser micromachining. *Sci. Rep.* **2015**, *5*, 8072. [CrossRef] [PubMed]
8. Kosma, K.; Zito, G.; Schuster, K.; Pissadakis, S. Whispering gallery mode microsphere resonator integrated inside a microstructured optical fiber. *Opt. Lett.* **2013**, *38*, 1301–1303. [CrossRef] [PubMed]
9. He, F.; Tang, J.; Song, J.; Lin, J.; Liao, Y.; Wang, Z.; Qiao, L.; Sugioka, K.; Cheng, Y. Fabrication of an integrated high-quality-factor (high-Q) optofluidic sensor by femtosecond laser micromachining. *Opt. Express* **2014**, *22*, 14792–14802.
10. Lahoz, F.; Martín, I.R.; Walo, D.; Freire, R.; Gil-Rostra, J.; Yubero, F.; Gonzalez-Elipe, A.R. Enhanced green fluorescent protein in optofluidic Fabry-Perot microcavity to detect laser induced temperature changes in a bacterial culture. *Appl. Phys. Lett.* **2017**, *111*, 111103. [CrossRef]
11. Floch, J.M.L.; Fan, Y.; Humbert, G.; Shan, Q.; Férachou, D.; Bara-Maillet, R.; Aubourg, M.; Hartnett, J.G.; Madrangeas, V.; Cros, D.; et al. Invited Article: Dielectric material characterization techniques and designs of high-Q resonators for applications from micro to millimeter-waves frequencies applicable at room and cryogenic temperatures. *Rev. Sci. Instrum.* **2014**, *85*, 489–493. [CrossRef] [PubMed]
12. Wu, X.; Oo, M.K.; Reddy, K.; Chen, Q.; Sun, Y.; Fan, X. Optofluidic laser for dual-mode sensitive biomolecular detection with a large dynamic range. *Nat. Commun.* **2014**, *5*, 3779. [CrossRef] [PubMed]

13. Gong, C.; Gong, Y.; Khaing Oo, M.K.; Wu, Y.; Rao, Y.; Tan, X.; Fan, X. Sensitive sulfide ion detection by optofluidic catalytic laser using horseradish peroxidase (HRP) enzyme. *Biosens. Bioelectron.* **2017**, *96*, 351–357. [CrossRef] [PubMed]

14. Chen, Q.; Liu, H.; Lee, W.; Sun, Y.; Zhu, D.; Pei, H.; Fan, C.; Fan, X. Self-assembled DNA tetrahedral optofluidic lasers with precise and tunable gain control. *Lab Chip* **2013**, *13*, 3351–3354. [CrossRef] [PubMed]

15. Özelci, E.; Aas, M.; Jonáš, A.; Kiraz, A. Optofluidic FRET microlasers based on surface-supported liquid microdroplets. *Laser Phys. Lett.* **2014**, *11*, 045802. [CrossRef]

16. Chen, Q.; Zhang, X.; Sun, Y.; Ritt, M.; Sivaramakrishnan, S.; Fan, X. Highly sensitive fluorescent protein FRET detection using optofluidic lasers. *Lab Chip* **2013**, *13*, 2679–2681. [CrossRef] [PubMed]

17. Aas, M.; Chen, Q.; Jonáš, A.; Kiraz, A.; Fan, X. Optofluidic FRET lasers and their applications in novel photonic devices and biochemical sensing. *IEEE J. Sel. Top. Quantum Electron.* **2015**, *22*, 188–202. [CrossRef]

18. Shopova, S.I.; Zhou, H.; Fan, X.; Zhang, P. Optofluidic ring resonator based dye laser. *Appl. Phys. Lett.* **2007**, *90*, 221101. [CrossRef]

19. Vollmer, F. Taking detection to the limit—Monitoring single molecule interactions on a label-free microcavity biosensor. *IEEE Photonics Technol. Lett.* **2014**, *28*, 4–10.

20. Baaske, M.D.; Foreman, M.R.; Vollmer, F. Single-molecule nucleic acid interactions monitored on a label-free microcavity biosensor platform. *Nat. Nanotechnol.* **2014**, *9*, 933–999. [CrossRef] [PubMed]

21. Wu, Y.; Zhang, D.Y.; Yin, P.; Vollmer, F. Ultraspecific and highly sensitive nucleic acid detection by integrating a DNA catalytic network with a label-free microcavity. *Small* **2014**, *10*, 2067–2076. [CrossRef] [PubMed]

22. Su, J. Label-free single molecule detection using microtoroid optical resonators. *J. Vis. Exp.* **2015**, *106*, e53180. [CrossRef] [PubMed]

23. Swaim, J.D.; Knittel, J.; Bowen, W.P. Detection of nanoparticles with a frequency locked whispering gallery mode microresonator. *Appl. Phys. Lett.* **2013**, *102*, 272–274. [CrossRef]

24. Zhu, J.; Özdemir, Ş.K.; He, L.; Chen, D.R.; Yang, L. Single virus and nanoparticle size spectrometry by whispering-gallery-mode microcavities. *Opt. Express* **2011**, *19*, 16195–16206. [CrossRef] [PubMed]

25. Li, M.; Wu, X.; Liu, L.; Fan, X.; Xu, L. Self-referencing optofluidic ring resonator sensor for highly sensitive biomolecular detection. *Anal. Chem.* **2013**, *85*, 9328–9332. [CrossRef] [PubMed]

26. Choi, C.J.; Belobraydich, A.R.; Chan, L.L.; Mathias, P.C.; Cunningham, B.T. Label-free optofluidic biosensing in microplate, microfluidic, and spot-based affinity capture assays. In Proceedings of the Lasers and Electro-Optics (CLEO) and Quantum Electronics and Laser Science Conference (QELS), San Jose, CA, USA, 16–21 May 2010; Volume 405, pp. 1–2.

27. Fan, X.; Yun, S.H. The potential of optofluidic biolasers. *Nat. Methods* **2014**, *11*, 141–147. [CrossRef] [PubMed]

28. Chen, Y.; Lei, L.; Zhang, K.; Shi, J.; Wang, L.; Li, H.; Zhang, X.M.; Wang, Y.; Chan, H.L.W. Optofluidic microcavities: Dye-lasers and biosensors. *Biomicrofluid* **2010**, *4*, 043002. [CrossRef] [PubMed]

29. Wang, W.; Zhou, C.; Zhang, T.; Chen, J.; Liu, S.; Fan, X. Optofluidic laser array based on stable high-Q Fabry-Pérot microcavities. *Lab Chip* **2015**, *15*, 3862–3869. [CrossRef] [PubMed]

30. Zhang, T.; Zhou, C.; Wang, W.; Chen, J. Generation of low-threshold optofluidic lasers in a stable Fabry-Pérot microcavity. *Opt. Laser Technol.* **2017**, *91*, 108–111. [CrossRef]

31. Lahoz, F.; Martín, I.R.; Gilrostra, J.; Olivaramirez, M.; Yubero, F.; Gonzalezelipe, A.R. Portable IR dye laser optofluidic microresonator as a temperature and chemical sensor. *Opt. Express* **2016**, *24*, 14383–14392. [CrossRef] [PubMed]

32. Lahoz, F.; Martín, I.R.; Walo, D.; Gil-Rostra, J.; Yubero, F.; Gonzalez-Elipe, A.R. A compact and portable optofluidic device for detection of liquid properties and label-free Sensing. *J. Phys. D Appl. Phys.* **2017**, *50*, 215103. [CrossRef]

33. Gerosa, R.M.; Sudirman, A.; Menezes, L.D.S.; Margulis, W.; Matos, C.J.S.D. All-fiber high repetition rate microfluidic dye laser. *Optica* **2015**, *2*, 186–193. [CrossRef]

34. Malak, M.; Pavy, N.; Marty, F.; Peter, Y.A.; Liu, A.Q.; Bourouina, T. Micromachined Fabry-Pérot resonator combining submillimeter cavity length and high quality factor. *Appl. Phys. Lett.* **2011**, *98*, 211113. [CrossRef]

35. Malak, M.; Gaber, N.; Marty, F.; Pavy, N.; Richalot, E.; Bourouina, T. Analysis of Fabry-Perot optical micro-cavities based on coating-free all-Silicon cylindrical Bragg reflectors. *Opt. Express* **2013**, *21*, 2378–2392. [CrossRef] [PubMed]

36. He, L.; Ouml, S.K.; Yang, L. Whispering gallery microcavity lasers. *Laser Photonics Rev.* **2013**, *7*, 60–82. [CrossRef]

37. Tu, X.; Wu, X.; Li, M.; Liu, L.Y.; Xu, L. Ultraviolet single-frequency coupled optofluidic ring resonator dye laser. *Opt. Express* **2012**, *20*, 19996–20001. [CrossRef] [PubMed]
38. Li, Z.L.; Zhou, W.Y.; Luo, M.M.; Liu, Y.G.; Tian, J.G. Tunable optofluidic microring laser based on a tapered hollow core microstructured optical fiber. *Opt. Express* **2015**, *23*, 10413–10420. [CrossRef] [PubMed]
39. Yu, J.; Liu, Y.; Luo, M.; Wang, Z.; Yang, G.; Zhang, H.W.; Zhang, X.H. Single longitudinal mode optofluidic microring laser based on a hollow-core microstructured optical fiber. *IEEE Photonics J.* **2017**, *9*, 7105510. [CrossRef]
40. Lee, W.; Kim, D.B.; Song, M.H.; Yoon, D.K. Optofluidic ring resonator laser with an edible liquid laser gain medium. *Opt. Express* **2017**, *25*, 14043–14048. [CrossRef] [PubMed]
41. Francois, A.; Riesen, N.; Gardner, K.; Monro, T.M.; Meldrum, A. Lasing of whispering gallery modes in optofluidic microcapillaries. *Opt. Express* **2016**, *24*, 12466–12477. [CrossRef] [PubMed]
42. Gu, F.; Xie, F.; Lin, X.; Linghu, S.; Fang, W.; Zeng, H.; Tong, L.; Zhuang, S. Single whispering-gallery mode lasing in polymer bottle microresonators via spatial pump engineering. *Light Sci. Appl.* **2017**, *6*, e17061. [CrossRef]
43. Xie, F.; Gu, F.; Wang, H.; Yao, N.; Zhuang, S.; Fang, W. Single-mode lasing via loss engineering in fiber-taper-coupled polymer bottle microresonators. *Photonics Res.* **2017**, *5*, B29–B33. [CrossRef]
44. Lu, Q.; Wu, X.; Liu, L.; Xu, L. Mode-selective lasing in high-Q polymer micro bottle resonators. *Opt. Express* **2015**, *23*, 22740–22745. [CrossRef] [PubMed]
45. Grimaldi, I.A.; Berneschi, S.; Testa, G.; Baldini, F.; Conti, G.N.; Bernini, R. Polymer based planar coupling of self-assembled bottle microresonators. *Appl. Phys. Lett.* **2014**, *105*, 2012–2016. [CrossRef]
46. Yang, Y.; Ward, J.; Nic Chormaic, S. Quasi-droplet microbubbles for high resolution sensing applications. *Opt. Express* **2014**, *22*, 6881–6898. [CrossRef] [PubMed]
47. Li, H.; Suter, J.D.; Reddy, K.; Lee, W.; Fan, X.; Sun, Y. Tunable single mode lasing from an on-chip optofluidic ring resonator laser. *Appl. Phys. Lett.* **2011**, *98*, 061103.
48. Chen, Q.; Ritt, M.; Sivaramakrishnan, S.; Sun, Y.; Fan, X. Optofluidic lasers with a single molecular layer of gain. *Lab Chip* **2014**, *14*, 4590–4595. [CrossRef] [PubMed]
49. Lee, W.; Chen, Q.; Fan, X.; Dong, K.Y. Digital DNA detection based on a compact optofluidic laser with ultra-low sample consumption. *Lab Chip* **2016**, *16*, 4770–4776. [CrossRef] [PubMed]
50. Chen, Y.C.; Chen, Q.; Fan, X. Optofluidic chlorophyll lasers. *Lab Chip* **2016**, *16*, 2228–2235. [CrossRef] [PubMed]
51. Yang, S.; Ta, V.D.; Wang, Y.; Chen, R.; He, T.; Demir, H.V.; Sun, H. Reconfigurable liquid whispering gallery mode microlasers. *Sci. Rep.* **2016**, *6*, 27200. [CrossRef] [PubMed]
52. Jiang, X.F.; Zou, C.L.; Wang, L.; Gong, Q.; Xiao, Y.F. Whispering-gallery microcavities with unidirectional laser emission. *Laser Photonics Rev.* **2016**, *10*, 40–61. [CrossRef]
53. François, A.; Riesen, N.; Ji, H.; Shahraam, A.V.; Monro, T.M. Polymer based whispering gallery mode laser for biosensing applications. *Appl. Phys. Lett.* **2015**, *106*, 60–82. [CrossRef]
54. Rui, C.; Ta, V.D.; Han, D.S. Single mode lasing from hybrid hemispherical microresonators. *Sci. Rep.* **2012**, *2*, 244.
55. Kryzhanovskaya, N.V.; Maximov, M.V.; Zhukov, A.E. Whispering-gallery mode microcavity quantum-dot lasers. *Quantum Electron.* **2014**, *44*, 189–200. [CrossRef]
56. Gu, F.; Zhang, L.; Zhu, Y.; Zeng, H. Free-space coupling of nanoantennas and whispering-gallery microcavities with narrowed linewidth and enhanced sensitivity. *Laser Photonics Rev.* **2015**, *9*, 682–688. [CrossRef]
57. Senthil, M.G.; Petrovich, M.N.; Jung, Y.; Wilkinson, J.S.; Zervas, M.N. Hollow-bottle optical microresonators. *Opt. Express* **2011**, *19*, 20773–20784. [CrossRef] [PubMed]
58. Aas, M.; Özelci, E.; Jonáš, A.; Fan, X. FRET lasing from self-assembled DNA tetrahedral nanostructures suspended in optofluidic droplet resonators. *Eur. Phys. J. Spec. Top.* **2014**, *223*, 2057–2062. [CrossRef]
59. Sun, Y.; Shopova, S.I.; Wu, C.S.; Arnold, S.; Fan, X.D. Bioinspired optofluidic fret lasers via DNA scaffolds. *Proc. Natl. Acad. Sci. USA* **2010**, *37*, 16039–16042. [CrossRef] [PubMed]
60. Testa, G.; Persichetti, G.; Bernini, R. Design and optimization of an optofluidic ring resonator based on liquid-core hybrid ARROWs. *IEEE Photonics J.* **2014**, *6*, 1–14. [CrossRef]
61. Khalil, D.; Dan, A. Volume refractometry of liquids using stable optofluidic Fabry-Pérot resonator with curved surfaces. *J. Micro Nanolithogr. MEMS MOEMS* **2015**, *14*, 045501.

62. Kushida, S.; Okada, D.; Sasaki, F.; Lin, Z.; Huang, J.; Yamamoto, Y. Lasers: Low-threshold whispering gallery mode lasing from self-assembled microspheres of single-sort conjugated polymers. *Adv. Opt. Mater.* **2017**, *5*, 1700123. [CrossRef]

63. Ren, L.; Zhang, X.; Guo, X.; Wang, H.; Wu, X. High-sensitivity optofluidic sensor based on coupled liquid-core laser. *IEEE Photonics Technol.* **2017**, *29*, 639–642. [CrossRef]

64. Ren, L.; Wu, X.; Li, M.; Zhang, X.; Liu, L.; Xu, L. Ultrasensitive label-free coupled optofluidic ring laser sensor. *Opt. Lett.* **2012**, *18*, 3873–3875. [CrossRef]

65. Zhang, X.; Ren, L.; Wu, X.; Li, H.; Liu, L.; Xu, L. Coupled optofluidic ring laser for ultrahigh-sensitive sensing. *Opt. Express* **2011**, *19*, 22242–22247. [CrossRef] [PubMed]

66. Reynolds, T.; Riesen, N.; Meldrum, A.; Fan, X.; Hall, J.M.; Monro, T.; Francois, A. Fluorescent and lasing whispering gallery mode microresonators for sensing applications. *Laser Photonics Rev.* **2017**, *11*, 1600265. [CrossRef]

67. Ciminelli, C.; Campanella, C.M.; Dell'Olio, F.; Campanella, C.E.; Armenise, M.N. Label-free optical resonant sensors for biochemical applications. *Prog. Quantum Electron.* **2013**, *37*, 51–107. [CrossRef]

68. Wang, Y.; Li, H.; Zhao, L.; Yang, J. A review of droplet resonators: Operation method and application. *Opt. Laser Technol.* **2016**, *86*, 61–68. [CrossRef]

69. Chistiakova, M.V.; Shi, C.; Armani, A.M. Label-free, single molecule resonant cavity detection: A double-blind experimental study. *Sensors* **2015**, *15*, 6324–6341. [CrossRef] [PubMed]

70. Ruan, Y.; Boyd, K.; Ji, H.; Francois, A.; Ebendorff-Heidepriem, H.; Munch, J.; Monro, T.M. Tellurite microspheres for nanoparticle sensing and novel light sources. *Opt. Express* **2014**, *22*, 11995–12006. [CrossRef] [PubMed]

71. He, L.; Ozdemir, S.K.; Zhu, J.; Kim, W.; Yang, L. Detecting single viruses and nanoparticles using whispering gallery microlasers. *Nat. Nanotechnol.* **2011**, *6*, 428–432. [CrossRef] [PubMed]

72. Zhu, J.; Zhong, Y.; Liu, H. Impact of nanoparticle-induced scattering of an azimuthally propagating mode on the resonance of whispering gallery microcavities. *Photonics Res.* **2017**, *5*, 396–406. [CrossRef]

73. Soria, S.; Berneschi, S.; Lunelli, L.; Nunzi Conti, G.; Pasquardini, L.; Pederzolli, C.; Righini, G.C. Whispering Gallery Mode Microresonators for Biosensing. *Adv. Sci. Technol.* **2013**, *82*, 55–63. [CrossRef]

74. Arnold, S.; Keng, D.; Shopova, S.I.; Holler, S.; Zurawsky, W.; Vollmer, F. Whispering gallery mode carousel—A photonic mechanism for enhanced nanoparticle detection in biosensing. *Opt. Express* **2009**, *17*, 6230–6238. [CrossRef] [PubMed]

75. Dantham, V.R.; Holler, S.; Barbre, C.; Keng, D.; Kolchenko, V.; Arnold, S. Label-free detection of single protein using a nanoplasmonic-photonic hybrid microcavity. *Nano Lett.* **2013**, *13*, 3347–3351. [CrossRef] [PubMed]

76. Zijlstra, P.; Paulo, P.M.; Orrit, M. Optical detection of single non-absorbing molecules using the surface plasmon resonance of a gold nanorod. *Nat. Nanotechnol.* **2012**, *7*, 379–382. [CrossRef] [PubMed]

77. Chan, J.; Thiessen, T.; Lane, S.; Gardner, K. Microfluidic detection of vitamin d3 compounds using a cylindrical optical microcavity. *IEEE Sens. J.* **2015**, *15*, 3467–3474. [CrossRef]

78. Zhu, H.; Dale, P.S.; Caldwell, C.W.; Fan, X. Rapid and label-free detection of breast cancer biomarker CA15-3 in clinical human serum samples with optofluidic ring resonator sensors. *Anal. Chem.* **2009**, *81*, 9858–9865. [CrossRef] [PubMed]

79. Suter, J.D.; Howard, D.J.; Fan, X. Label-free DNA methylation analysis using the optofluidic ring resonator sensor. *Biosens. Bioelectron.* **2015**, *26*, 1016–1020. [CrossRef] [PubMed]

80. Zhou, Z.H.; Shu, F.J.; Zhen, S.; Dong, C.H.; Guo, G.C. High-Q whispering gallery modes in a polymer microresonator with broad strain tuning. *Sci. China Phys. Mech. Astron.* **2015**, *58*, 1–5. [CrossRef]

81. Chang, L.; Jiang, X.S.; Hua, S.Y.; Yang, C.; Wen, J.M.; Jiang, L.; Li, G.Y.; Wang, G.Z.; Xiao, M. Parity-time symmetry and variable optical isolation in active-passive-coupled microresonators. *Nat. Photonics* **2014**, *8*, 524–529. [CrossRef]

82. Ward, J.; Benson, O. WGM microresonators: Sensing, lasing and fundamental optics with microspheres. *Laser Photonics Rev.* **2011**, *5*, 553–570. [CrossRef]

83. Gilardi, G.; Beccherelli, R. Integrated optics nano-opto-fluidic sensor based on whispering gallery modes for picoliter volume refractometry. *J. Phys. D Appl. Phys.* **2013**, *46*, 1–9. [CrossRef]

84. Giorgini, A.; Avino, S.; Malara, P.; Natale, P.D.; Gagliardi, G. Fundamental limits in high-Q droplet microresonators. *Sci. Rep.* **2017**, *7*, 41997. [CrossRef] [PubMed]

85. Zhang, M.; Liu, B.; Wu, G.; Chen, D. Hybrid plasmonic microcavity with an air-filled gap for sensing applications. *Opt. Commun.* **2016**, *380*, 6–9. [CrossRef]

86. Arbabi, E.; Kamali, S.M.; Arnold, S.; Goddard, L.L. Hybrid whispering gallery mode/plasmonic chain ring resonators for biosensing. *Appl. Phys. Lett.* **2014**, *105*, 231107. [CrossRef]

87. Nadgaran, H.; Garaei, M.A. Enhancement of a whispering gallery mode microtoroid resonator by plasmonic triangular gold nanoprism for label-free biosensor applications. *J. Appl. Phys.* **2015**, *118*, 043101. [CrossRef]

88. Xiao, Y.F.; Li, B.B.; Jiang, X.; Hu, X.; Li, Y.; Gong, Q. High quality factor, small mode volume, ring-type plasmonic microresonator on a silver chip. *J. Phys. B At. Mol. Opt.* **2010**, *43*, 035402. [CrossRef]

89. Bozzola, A.; Perotto, S.; De, A.F. Hybrid plasmonic-photonic whispering gallery mode resonators for sensing: A critical review. *Analyst* **2017**, *142*, 883–898. [CrossRef] [PubMed]

90. Santiagocordoba, M.A.; Boriskina, S.V.; Vollmer, F.; Demirel, M.C. Nanoparticle-based protein detection by optical shift of a resonant microcavity. *Appl. Phys. Lett.* **2011**, *99*, 073701. [CrossRef]

91. Zhang, X.; Liu, L.; Xu, L. Ultralow sensing limit in optofluidic micro-bottle resonator biosensor by self-referenced differential-mode detection scheme. *Appl. Phys. Lett.* **2014**, *104*, 033703. [CrossRef]

92. Tang, T.; Wu, X.; Liu, L.; Xu, L. Packaged optofluidic microbubble resonators for optical sensing. *Appl. Opt.* **2016**, *55*, 395–399. [CrossRef] [PubMed]

93. Shang, J.; Dai, H.; Zou, Y.; Chen, X. Detection of low-concentration EGFR with a highly sensitive optofluidic resonator. *Chin. Opt. Lett.* **2017**, *15*, 092301. [CrossRef]

94. Su, J.; Goldberg, A.F.; Stoltz, B.M. Label-free detection of single nanoparticles and biological molecules using microtoroid optical resonators. *Light Sci. Appl.* **2017**, *5*, e16001. [CrossRef]

95. Lin, G.; Chembo, Y.K. Phase-locking transition in Raman combs generated with whispering gallery mode resonators. *Opt. Lett.* **2016**, *41*, 3718–3721. [CrossRef] [PubMed]

96. Deych, L.; Shuvayev, V. Theory of nanoparticle-induced frequency shifts of whispering-gallery-mode resonances in spheroidal optical resonators. *Phys. Rev. A* **2015**, *5*, e625–e626. [CrossRef]

97. Bog, U.; Laue, T.; Grossmann, T.; Beck, T.; Wienhold, T.; Richter, B.; Hirtz, M.; Fuchs, H.; Kal, H.; Mappes, T. On-chip microlasers for biomolecular detection via highly localized deposition of a multifunctional phospholipid ink. *Lab Chip* **2013**, *13*, 2701–2707. [CrossRef] [PubMed]

98. Li, Q.; Eftekhar, A.A.; Xia, Z.; Adibi, A. A unified approach to mode splitting and scattering loss in high-Q whispering-gallery-mode microresonators. *Phys. Rev. A* **2013**, *88*, 8981–8995. [CrossRef]

99. Scholten, K.; Collin, W.R.; Fan, X.; Zellers, E.T. Nanoparticle-coated micro-optofluidic ring resonator as a detector for microscale gas chromatographic vapor analysis. *Nanoscale* **2015**, *7*, 9282–9289. [CrossRef] [PubMed]

100. Collin, W.R.; Scholten, K.W.; Fan, X.; Paul, D.; Kurabayashi, K.; Zellers, E.T. Polymer-coated micro-optofluidic ring resonator detector for a comprehensive two-dimensional gas chromatographic microsystem: µGC × µGC − µOFRR. *Analyst* **2016**, *141*, 261–269. [CrossRef] [PubMed]

101. Madani, A.; Harazim, S.M.; Bolaños Quiñones, V.A.; Kleinert, M.; Finn, A.; Ghareh Naz, E.S.; Ma, L.; Schmidt, O.G. Optical microtube cavities monolithically integrated on photonic chips for optofluidic sensing. *Opt. Lett.* **2017**, *42*, 486–489. [CrossRef] [PubMed]

102. Smith, E.J.; Xi, W.; Makarov, D.; Mönch, I.; Harazim, S.; Bolaños Quiñones, V.A.; Schmidt, C.K.; Mei, Y.; Sanchez, S.; Schmidt, O.G. Lab-in-a-tube: Ultracompact components for on-chip capture and detection of individual micro-/nanoorganisms. *Lab Chip* **2012**, *12*, 1917–1931. [CrossRef] [PubMed]

103. Harazim, S.M.; Bolaños Quiñones, V.A.; Kiravittaya, S.; Sanchez, S.; Schmidt, O.G. Lab-in-a-tube: On-chip integration of glass optofluidic ring resonators for label-free sensing applications. *Lab Chip* **2012**, *12*, 2649–2655. [CrossRef] [PubMed]

Review

Optofluidics in Microstructured Optical Fibers

Liyang Shao [1], Zhengyong Liu [2],*, Jie Hu [1], Dinusha Gunawardena [2] and Hwa-Yaw Tam [2]

[1] Department of Electrical and Electronic Engineering, Southern University of Science and Technology, Shenzhen 518055, China; shaoly@sustc.edu.cn (L.S.); s1604080303@stu.cjlu.edu.cn (J.H.)

[2] Photonics Research Center, Department of Electrical Engineering, The Hong Kong Polytechnic University, Kowloon, Hong Kong; dinusha.gunawardena@polyu.edu.hk (D.G.); eehytam@polyu.edu.hk (H.-Y.T.)

* Correspondence: zhengyong.liu@connect.polyu.hk

Received: 27 January 2018; Accepted: 21 March 2018; Published: 24 March 2018

Abstract: In this paper, we review the development and applications of optofluidics investigated based on the platform of microstructured optical fibers (MOFs) that have miniature air channels along the light propagating direction. The flexibility of the customizable air channels of MOFs provides enough space to implement light-matter interaction, as fluids and light can be guided simultaneously along a single strand of fiber. Different techniques employed to achieve the fluidic inlet/outlet as well as different applications for biochemical analysis are presented. This kind of miniature platform based on MOFs is easy to fabricate, free of lithography, and only needs a tiny volume of the sample. Compared to optofluidics on the chip, no additional waveguide is necessary to guide the light since the core is already designed in MOFs. The measurements of flow rate, refractive index of the filled fluids, and chemical reactions can be carried out based on this platform. Furthermore, it can also demonstrate some physical phenomena. Such devices show good potential and prospects for applications in bio-detection as well as material analysis.

Keywords: optofluidics; microstructured optical fiber; MOF; sensors

---

1. Introduction

Optofluidics has become more and more attractive and promising in the recent decade due to its features of small consumption of volume, fast response, and high sensitivity, as well as its capability of interacting light with substance, especially with chemical solutions [1]. Basically, it combines the disciplines of microfluidics and waveguides so that light-matter interaction can occur in direct or indirect ways. Typically, direct interaction means that the liquid flows in the channel also act as the optical core (e.g., hollow-core waveguide [2,3]), whereas indirect interaction means that liquids and light are separated but mutually influenced via indirect means (e.g., evanescent field [4], surface plasmonic resonance [5], heat flux [6], mechanic force [7]). Many interesting applications of optofluidic devices have been investigated, such as lab-on-chip and lab-in-fiber devices. Through the use of such technology, it is possible to easily achieve live cell imaging, chemical synthesis, particle trapping/sorting, biological analysis, high performance sensors (for e.g., DNA, refractive index (RI), flow rate), etc. [8]. However, most of these applications are carried out on a chip with customized microchannels. These channels vary from tens to hundreds of micrometers to allow for different analytes or particles in solution to pass through and eventually be analyzed. Microfluidics has blossomed extensively thanks to lab-on-chip technology [9]. To fabricate the tiny microchannels on a chip made of, e.g., SU-8, PDMS, silica, etc., micro/nanofabrication technologies such as photo- or soft-lithography are necessary, which, to some extent, increases the fabrication complexity and cost. Nevertheless, microfluidics provides a good platform for biological and chemical analysis, especially after being integrated with optics, which has occurred over the past 10 years.

In addition to lab-on-chip technology being an excellent platform to conduct microfluidics due to the flexibility of designing microchannels, specialty optical fiber is another promising option to combine microfluidics with optics owing to the light guidance in the fiber, which is thus called lab-in-fiber [10]. By microfabricating structures on the fiber facet where the light meets measurand, the refractive index of a specific liquid can be easily characterized using the hybrid metallo-dielectric nanoprobes [11] or Fabry–Pérot (FP) cavity [12]. As reported recently in 2017, a microbubble was also fabricated on the fiber end facet to realize an optofluidic interferometer [13]. Apart from the microstructures on the fiber tip interacting with liquids, the gap between two right-cleaved optical fibers is employed to conduct optofluidic measurement as well. Typically, such a gap is part of the microchannel where the target liquid flows. Two fiber Bragg gratings (FBGs) inscribed on these two fibers form an FP cavity that covers the gap so that the change of the fluids (e.g., RI, flow rate) can be detected by the optical resonance signal [14,15]. Furthermore, Leite et al. reported in 2017 that the light coming out from the end facet of a high NA (numerical aperture) multimode fiber can trap the particle tightly and move it in three dimensions [16], which provides the technology to manipulate particles or cells in optofluidics.

Although the fiber end facet is a good place for light to come out and interact with fluids directly, the contact area is still limited. The advent of microstructured optical fibers (MOFs), especially hollow-core photonic crystal fiber (PCF), offers a perfect platform for guiding the fluids and light on the same path and along a longer length to make direct/indirect interaction possible. What is more, MCF fibers made of silica are more resistant to external interference compared to plastic microchips due to the properties of silica, such as being inert, having low auto-fluorescence, being non-deformable under high pressure, and so on. Microstructured fiber can also be suitable for remote sensing because of the fact that MCF can be fabricated in kilometer lengths at low cost. The air hole region acts as a place for chemical reaction under the core light irradiation, which is very promising in the applications of chemical sensing and photochemistry [17]. Using such a technique, the process of catalytic reaction could be analyzed in ultralow concentration due to the tiny amount of sample filled in the core [18], which, in contrast, is not easy to achieve in a conventional photoreactor system. Under the irradiation of the light guided in the hollow core, the conversion from cyanocobalamin (CNCbl) to aquacobalamin (H_2OCbl) was demonstrated by Unterkofler et al., which is comparable to the reaction in the cuvette. Moreover, one study reported that the qualitative detection of Cy-5-labeled DNA molecules in aqueous solution is made possible by filling a PCF with the sample and measuring the transmission absorption [19]. Additionally, a good review on the development of optofluidic PCF-based sensors was recently put forth by Ertman et al., in which RI, mechanical quantities, electrical field, and magnetic field sensors employing optofluidics in PCFs are reviewed [20]. Overall, interest has increased significantly in the applications of optofluidics in MOFs.

In this paper, we give a short review of the optofluidics based on the platform of MOFs, especially focusing on the techniques used to implement optofluidic sensors and their applications. In next section, a brief introduction of MOFs together with the technical issues of inlet/outlet are given, followed by the encouraging applications realized by MOF optofluidics. The conclusion is presented at the end.

2. Implementation of Optofluidics Based on Microstructured Optical Fibers

Since the first PCF was invented and fabricated by Phillip Russel et al. in the 1990s, they have attracted increasing attention from researchers in various disciplines [21]. Many fiber devices based on PCFs have been developed. MOF is such a specialty optical fiber that possesses air holes arranged in a certain structure along the fiber length, and PCF is a type of MOF that consists of a honeycomb structure with air holes compared to silica glass, as shown in Figure 1a–c. It is worth noting that the structure of MOFs can also be realized by using lower-index rods instead of air holes; however, this structure is not suitable for the use of optofluidics as no air channels exist in the fiber. To achieve optofluidics where microchannels are needed, fibers with an air hole structure are desirable. MOFs are more general and have more flexibility in fabrication as long as the structure is able to confine light. The structure of an MOF is not limited to the periodic structure arrangement, which makes it possible to maintain a few large air holes in the cladding [22,23] while suspending a small core in the center, as shown in Figure 1d–g. Therefore, these MOFs can be considered as a good option when employing evanescent field as an indirect interaction in optofluidics. In particular, there are two types of MOFs. One consists of a solid core, whereas the other consists of a hollow core. Typically, the former follows the light-guiding mechanism of total internal reflection similar to conventional optical fibers, whereas the latter MOFs obey the theories of either photonic bandgap (PBG) [24] or anti-resonant reflecting optical waveguide (ARROW) [25].

Particularly, the single-mode fiber consists of a core and cladding, which have a high and relatively low refractive index, respectively, and the light is guided based on total internal reflection (TIR). The MOFs with a solid core follow a similar guiding mechanism, in which the average index of the cladding is lower than that of the solid core due to the existing air holes. Therefore, it can also be called a modified TIR (M-TIR), and the MOFs obeying this principle can be called index-guiding MOFs. In terms of the hollow-core MOFs, the core is completely air, and its index close to 1, much lower than that of glass. In this case, TIR is not satisfied. However, owing to the periodic structure consisting of a low index part (e.g., air holes) and a high index part (e.g., glass), it has a Bragg reflection effect, which leads to several photonic bandgaps that allow light to be confined in the core. The certain wavelengths of the light leaking into the cladding can be reflected back into the core. Basically, the liquids filled in the hollow core have lower index than glass, thus the photonic bandgap should be applied as well to demonstrate optofluidics in this kind of MOF.

The index-guiding MOFs have air holes arranged in the cladding, which can be fully or selectively filled with an aqueous solution. When the core is reduced to 1–2 µm, the evanescent field is strong enough to penetrate into the air holes and interact with the fluids filled inside. On the other hand, the average refractive index of the air holes region changes with the filled fluids, which in turn influences the propagation constant of the core mode. For example, the RI sensitivity can be up to 38,000 nm/RIU (refractive index unit) when selective air holes are filled with liquid, resulting in a detection limit of 4.6×10^{-7} [26]. However, the light-matter interaction is more straightforward for the hollow-core MOFs as the light and fluids propagate in the core simultaneously. Most PCF-based optofluidics for photochemistry use hollow-core MOFs [17]. A liquid-core fiber proposes another interesting application, where the specific fluids infiltrated in the center hole become a new optical core and the optical properties of these liquids (e.g., nonlinearity) are employed to achieve photonic phenomena such as hybrid solitons and supercontinuum generation [27]. Hence, the flexibility of design as well as the fabrication makes MOFs very promising and attractive for optofluidics.

Figure 1 summarizes the state-of-the-art implementations of optofluidics based on various MOFs and the possible strategies that have been adopted to fabricate the inlet/outlet for the aqueous solution. The air holes in the cladding of the PCFs can be fully [28] or partially [29] filled with liquids to realize high sensitive RI sensors. Basically, the liquids with low viscosity can flow into the air channels automatically due to the effect of capillary force. However, an additional pump may be needed for highly viscous fluids [3]. Index-guiding PCFs as well as hollow-core PCFs are two main fibers used to conduct optofluidics, and the latter has been demonstrated to function excellently in photochemistry. In 2017, Gao et al. successfully demonstrated that in an anti-resonant MOF (e.g., Figure 1e), where light is guided based on the principle of ARROW, the RI and flow rate can be measured simultaneously [30]. This provides a good paradigm to carry out optofluidics in such a type of fiber. Moreover, the flexibility of fabricating MOFs allows the existence of large air holes, as shown in Figure 1, which makes it easier for liquids to enter the fiber. The tiny core of the fiber can be suspended in the center by three, four, or six large air holes [4], as shown in Figure 1d, or attached to the inner surface of a capillary [31], as shown in Figure 1f. The light-matter interaction of such fibers with large holes occurs due to the strong evanescent field from the tiny core.

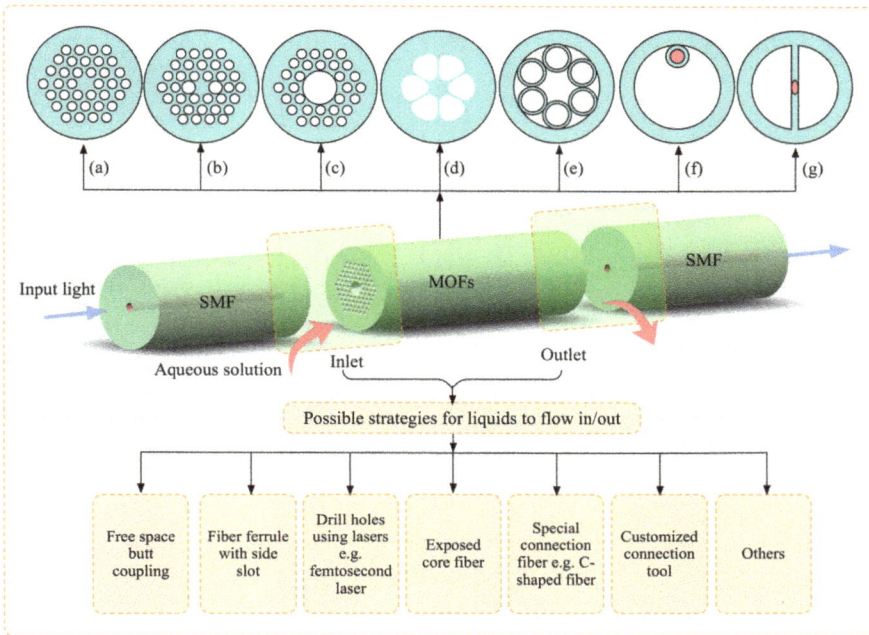

Figure 1. Schematic figure of the optofluidics based on various microstructured optical fibers (MOFs).

To achieve optofluidics in MOFs, regardless of which type of fiber is employed, it is technically challenging to fill the aqueous solution into the air holes of MOFs properly, especially for solutions with high viscosity. There have been some excellent techniques reported to enable liquids to flow into the air holes of MOFs [2,4,15,17,28,29,32–34]. The initial and easy way is to butt-couple light into the MOFs, leaving a gap between the SMF (single mode fiber) and MOFs so that fluids can enter the air holes [2]. However, this requires very careful alignment and additional precise XYZ stages. Alternatively, the idea of inserting two fibers into the ferrule with a side slot has been realized, where the fluids enter via the open slot [35]. The free-spacing coupling has been found to be a convenient strategy because no additional special fibers or structures are needed. Another straightforward way is to directly drill entrance holes in the outer surface of MOFs, which can open specific inner holes of MOFs to external access [30]. In this way, the MOFs can be spliced to SMFs easily in advance; however, this method relies on high power laser technology such as femtosecond laser micromachining. In terms of the suspended-core MOFs, T.M. Monro et al. pioneered the utilization of exposed-core fiber to conduct optofluidics, where one of the three holes in the fiber is mechanically polished and opened during the preparation of the fiber [4]. Eventually the tiny fiber core is exposed and can interact with and sense the materials based on fluorescence. In 2013, Wu et al. adopted a brilliant way of connecting a short C-shaped fiber between an SMF and PCF to measure the RI, where the water can enter the air holes of the PCF via the open C-shaped fiber [28]. This technology allows the MOF-based optofluidic device to be assembled using only a procedure of splicing. In addition to these special fibers, other customized connection tools have been developed as well to provide entrance for the fluids, including microfluidic chips capable of mounting fibers [3] and capillary-assisted arrangements [34]. Similar to the lab-on-chip setup, a syringe pump can be utilized for all of these strategies to improve the filling efficiency, especially when the fluids have relatively high viscosity.

The flexibility of MOFs fabrication as well as the inlet/outlet design makes the optofluidics implemented in MOFs quite promising and encouraging. No additional waveguide for light is needed. Since all of the test configurations are fiber-based, such optofluidic devices could be very miniature. The mature fiber-optic sensing technologies, e.g., grating techniques (i.e., fiber Bragg grating (FBG), long period grating (LPG), tilted fiber Bragg grating (TFBG), etc.), as well as interferometry (i.e., Fabry–Pérot interferometry, Sagnac interferometry, Mach-Zehnder interferometry, etc.), can be used. Moreover, fluoroscopy and the absorption of materials have been widely utilized to sense analytes filled in MOFs. It can be anticipated that, together with lab-on-chip microfluidics, the platform of MOFs will play an increasingly important role in optofluidics.

3. Applications

3.1. Microfluidic Sensing Mechanism

According to the classification of sensing mechanisms, like other optical fiber sensors, microstructured optical fiber sensors can be broadly divided into light intensity absorption-based sensors, wavelength-shift-based sensors, Stimulated Raman-scattering-based sensors, and fluorescence-based sensors.

Light intensity absorption-based detection is one of the most commonly used detection methods. It is well known that the evanescent energy distributed in the microfluidic channel of the microstructured fiber would be absorbed by the analyte, which influences the intensity of the transmitted light. This sensing mechanism is typical of fiber-based chemical sensors and traditional analytical instruments. Although different wavelengths can be chosen, UV-Vis-NIR absorption is one of the most commonly used light-based methods for the chemical identification of analytes because many analytes can be detected by their absorption changes in this region. For instance, Mona Nissen et al. introduced the concept of UV spectroscopy in liquid-filled anti-resonant hollow-core fibers to carry out the detection of pharmaceuticals, as shown in Figure 2. The limit of detection (LOD) for sulfamethoxazole (SMX) and sodium salicylate (SS) reached down to 0.1 μM (26 ppb) and 0.4 μM (64 ppb), respectively [36].

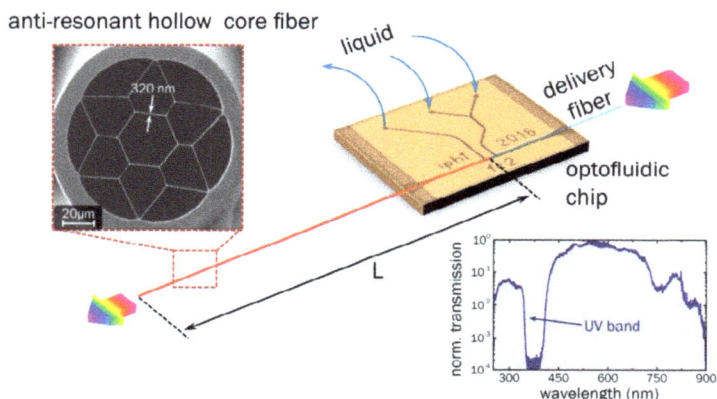

Figure 2. Schematic of the sensor for pharmaceuticals based on UV spectroscopy and water-filled anti-resonant hollow-core fibers (red), as well as the microfluidic chip (orange) and the delivery fiber (light blue), which are proposed in Reference [36]. The two insets show a scanning electron micrograph image of the cross-section of the silica microstructured fiber (upper left, the number refers to the average strand thickness in nm) and an example sample transmission spectrum (normalized to the maximum transmission value at 560 nm) where the fiber is filled with water (lower right).

It is already a well-known mechanism to detect the shift of the transmission spectrum due to the refractive index variation of the analyte. Despite the relatively good sensitivity, there are many interesting ways to further improve its performance. Surface plasmon resonance (SPR) is the most well-known method. A wavelength-shift-based sensor is linked to refractive index sensing, and will be covered in more detail in the next section.

Raman spectroscopy is a powerful tool for determining molecular structure and has been widely used in many aspects. For example, V. S. Tiwari et al. monitored the concentration ratio of liquid oxygen and nitrogen in a cryogenic mixture based on a fiber-optic Raman sensor [37]. However, the Raman scattering is extremely weak, and thus a high-sensitivity spectrometer and a high-power pump source are necessary. Obtaining an adequate Raman signal of analyte molecules in an optical fiber has become a hot topic [38,39]. F. M. Cox et al. first demonstrated that surface-enhanced Raman scattering (SERS) signals can be enhanced by the guidance in a hollow-core MOF [40]. A. Amezcua-Correa et al. proposed a SERS probe based on an MOF substrate, where the metal nanoparticles were deposited on the inner surface of the air channels of the MOF [41]. The next year, H. Yan et al. designed surface-enhanced Raman probe based on an index-guided PCF, and compared the performance of SERS obtained when the nanoparticles were coated on the inner surface of the air holes and mixed in the analyte solution. The experimental results show that by mixing the gold nanoparticles in the solution, the sensor can attach more analyte molecules, thus leading to a higher sensitivity [42]. G. Wang et al. explored gas Raman sensing in a multi-opened-up suspended-core fiber, where the multi-opened-up structure accelerated the gases diffusion. As a result, real-time Raman sensing could be realized with a response time of less than 6 s [43].

Fluorescence is the photoluminescence of cold luminescence, i.e., the light of a longer wavelength emitted by a substance that absorbs light or other electromagnetic radiation, which has been widely used in microscopy, environmental pollutant monitoring, and biosensing for many years. Some scholars use optical fibers with a tiny core to fabricate sensors, so that there is a large amount of evanescent field energy interacting with the analyte, leading to light absorption or a fluorescence effect that can be utilized for sensing. Optofluidics in MOFs integrate the light path and fluid together into a single strand of fiber so as to increase the interaction length between the light field and analyte, which eventually can improve the sensitivity of fluorescence-based detection [44,45]. S. C. Warren-Smith et al. presented two new methods to fabricate exposed-core fiber and demonstrated that, compared with traditional MOFs with a protected core [4], such exposed-core MOFs can achieve a shorter response time in fluorescence detection. It is worth noting that not all substances can produce fluorescence. In addition to the fact that those limited types of substances that produce fluorescence are often used as fluorescent markers to detect other materials, fluorescence detection is often associated with biochemical sensing or other molecular detection [46,47].

3.2. Microfluidic Detection Applications

3.2.1. Physical Parameters Detection inside Fluids

Microfluidic Flow Rate Detection

The microfluidic flow rate is a critical parameter in the industrial, chemical, and biological realms. Precise control of the flow rate has been widely used in nanotechnology and cell biology. Traditional flow rate optical fiber sensors are mostly based on the lateral force of a fluid acting on the surface of the fiber probe, making the fiber optic probe bend, causing the strength or phase of the optical signal transmitted within the fiber to change, and thus achieving the purpose of detection [48,49]. With the popularity of lab-on-chip designs, many teams have integrated fiber sensing probes with polymers into a single chip [7,50]. The microfluidic channel for liquid flow crosses the fiber channel. The chip integration not only reduces the size of the sensor but also reduces the amount of sample required. Microstructured fibers combine the optical path with the liquid flow path inside the fiber, which greatly increases the sensor's integration. Such fluid flow in this way will not bend the fiber. However, fluid flow can take away heat. P. Christodoulides's research group explored the relationship between liquid flow and heat transfer in the fiber [51]. Zhimei Qi's research group proposed an anti-resonant reflecting guidance, a hollow-core photonic crystal fiber coated with a few layers of grapheme on the surface, and realized the simultaneous measurement of flow rate and refractive index [30]. The graphene film grown using the chemical vapor deposition technique was rinsed in deionized water several times and floated on the water surface. The graphene was rinsed into the hollow-core photonic crystal fiber with water, and coated on the surface of the hollow-core after the water was dried. Figure 3 shows the experimental setup of this sensor. A 532-nm laser connected with a beam expander was used to illuminate the few layers of grapheme in the hollow-core PCF through the bottom hollow slot of the metal plate. As the fluid flows at different velocities in the microfluidic channels in the fiber, the heat transfer rates and refractive index of the grapheme layers are also different, resulting in different transmission intensities. The experimental results show that a sensitivity of up to -2.99 dB/(μL/s) was achieved, and could be further improved by increasing the heating intensity.

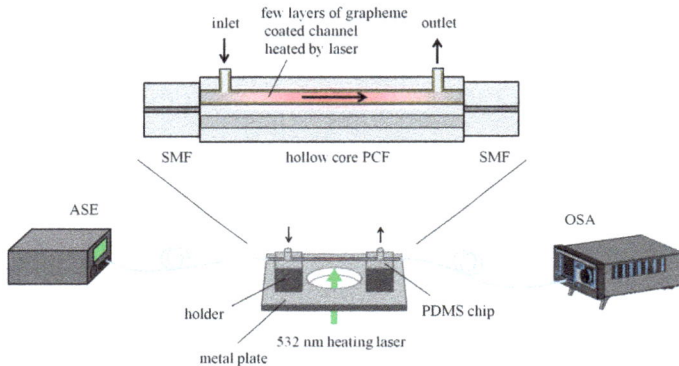

Figure 3. The schematic of the experimental setup for flow rate detection in Reference [30].

Refractive Index Measurement

Refractive index is a basic attribute of matter, indicating the ratio of the velocity of light in vacuum to the speed in this optical medium. Refractive index sensing is always an interesting application, and there is no exception within the optofluidics in microstructured optical fibers.

G. Wang et al. incorporated a side-opened fluidic microchannel and suspended-core structure into a dual-core PCF, and explored RI detection based on the dual-beam interference mechanism. The phase shift from the change of the birefringence, which corresponds to the refractive index of the fluidic analyte, was calculated. As a result, the sensor showed a good sensitivity and the LOD reached 2.2×10^{-6} RIU [52].

The optical leaky loss is induced due to the slight difference of refractive index between the analyte and the optical fiber through the evanescent field interaction. G. Wang et al. designed a side-opened surface plasmon resonance (SPR) sensor based on a suspended-core fiber where the core is exposed and coated by an Au film. They detected the transmitted light intensity and the results show that the LOD was 2.3×10^{-5} RIU [53].

Because the change of RI of fluid analyte in the MOFs gives rise to the variation of optical path, most RI sensors detect the wavelength shift of characteristic spectral peaks or valleys with the variation of the RI.

G. Bertrand et al. designed a solid-core PCF-based SPR sensor. In this sensor, two large channels for liquid analyte in the fiber were plated with gold to excite the SPR mode and improve the sensitivity;

the detection resolution could reach 7.2×10^{-6} RIU [54]. The RI sensor based on twin-core photonic bandgap fibers proposed by Y. Wu et al. could also reach a detection limit of 10^{-6} RIU at wavelength 1100 nm [55]. X. Yu et al. designed an RI sensor by writing a long-period grating in a PCF, and measured an RI ranging from 1.32 to 1.39. They obtained a sensitivity of 4.1×10^{-6} RIU [56]. K. C. Darran filled the liquid into selective air holes of a PCF, and found that the maximum sensitivity of this RI sensor could reach up to 38,000 nm/RIU, as well as achieve a detection limit of 4.6×10^{-7} RIU [26]. Multichannel sensing based on the Fano resonance of the whispering gallery modes in MOFs was also proposed with the sensitivity of tens of nm/RIU in TE (Transverse Electrical) mode or TM (Transverse Magnetic) mode [57]. Furthermore, the simultaneous measurements of refractive index and other parameters are demonstrated in References [30,58]. We summarized the detection performance of those sensors in Table 1.

Table 1. Performance of refractive index (RI) sensors based on optofluidics in microstructured optical fibers.

Structure	Sensitivity	Limit of Detection	Reference/Year
Side-opened and suspended-dual-core fiber	8360 rad/RIU*	2.2×10^{-6} RIU	[52]/2012
Side-opened suspended-core fiber with surface plasmon resonance (SPR)	3500 nm/RIU	2.3×10^{-5} RIU	[53]/2015
Photonic bandgap fiber-with SPR	13,750 nm/RIU	7.2×10^{-6} RIU	[54]/2007
Twin-core all-solid photonic bandgap fibers		10^{-6} RIU	[55]/2009
Single-mode photonic crystal fiber (PCF) with long-period grating	243 nm/RIU	4.1×10^{-6} RIU	[56]/2008
Photonic crystal fiber with selective hole(s) filled by fluid	30,100 nm/RIU	4.6×10^{-7} RIU	[26]/2009
Exposed-core MOF with SPR	1900 nm/RIU~12,500 nm/RIU		[57]/2016
Graphene coated hollow-core PCF	1328 nm/RIU		[30]/2017
Capillary channels (different diameters) inside a tubular frame	13.3181 nm/RIU~29.0557 nm/RIU		[58]/2017

*RIU: refractive index unit.

3.2.2. Chemosensing

PH Sensing

PH is a critical parameter of fluids, describing the degree of acid-base strength of aqueous solution, and the detection of pH is of great importance in industrial and chemical fields. F. M. Cox et al. drilled a lateral hole into one of the air channels of polymer optical fibers and used it for chemical sensing. Owing to the easy interaction between the side-opened core and the ambient analyte, this chemical sensor can realize real-time measurement. Bromothymol blue (BTB) was used as an indicator to test the sensing characteristics of a slotted microstructured polymer optical fiber, and the change in color of the BTB was used to show the change in the pH of the solution inside the fiber. The acidic solution and the basic solution corresponded to completely different spectra, although this was only a qualitative result. The experimental results demonstrated that the fiber can serve as a real-time evanescent wave absorption spectroscopy pH sensor using bromothymol blue as an indicator [59].

X. H. Yang et al. designed a pH sensor based on a polymer (PMMA) MOF, where the air channels were deposited by a cellulose acetate (CA) thin film doped with pH-sensitive fluorescence dye (eosin). Dynamic tests showed that the eosin-CA-MPOF (microstructured polymer optical fiber) probe can respond quickly, within 1 s, due to the nanoscale thickness of the membrane and its good hydrophilicity. Fluorescence intensity showed high stability at each pH value. The measured pH value ranged from pH 2.5 to 4.5, and could be further expanded from 1.5 to 4.5 if using surfactants hexadecyl trimethyl ammonium bromide [60].

Gas Monitoring

With the improvement of people's awareness of environmental protection and the consequent strict standards for gas emissions, it is increasingly important to develop sensitive and efficient gas detection methods. Optofluidics in MOFs provide a potential method to develop a sensitive, real-time gas sensor with high integration. Experiments on PCF-based gas detectors were initially used with evanescent wave sensing in a PCF [61–63]. However, in this type of optical fiber sensor, the sensitivity obtained was very low due to the low overlap between the gas sample and the optical field energy and their weak interaction. In a pioneering experiment, T. Ritari et al. demonstrated gas (methane and ammonia) detection in a hollow-core PCF by observing the direct absorption spectra in the near-infrared band [64]. Using similar techniques, hollow-core PCFs have since been used in detecting acetylene (C_2H_2) [65] and ammonia (NH_3) at 1.5 μm [66], and methane (CH4) at 3.3 μm [67].

For a few meters of fiber length, it may take several hours to fill by the free diffusion of gas in the air channels of the PCF. An important challenge for PCF gas detection is the ability to achieve a rapid response to sudden changes and real-time monitoring. It has recently been shown that lateral access holes can be made in the PCF cladding, allowing the gas to reach the hollow faster. Y. L. Hoo et al. cut out a 7-cm long piece of a hollow-core PCF, drilled seven mini-holes at the side of the PCF, and formed seven microchannels and the cross-section of a microchannel. A mixture of methane and nitrogen was inserted into the chamber of the hollow-core photonic bandgap fiber with a slow rate through these seven channels at the side, and the methane was detected with an LOD of ~647 ppm within a ~3-s response time [68]. S. H. Kassani et al. introduced a C-shaped fiber between multimode fibers and a suspended ring-core PCF, which made it easier for fluids to flow into the air channels of the PCF. In such a way, the sensitivity was improved and the response time was shortened [69]. X. Zhou et al. wrote a spiral trench on the surface of an SMF and wrote an FBG in the fiber, and then Pt-WO_3 films, sensitive to hydrogen, were modified on the surface of the fiber. As a result, a hydrogen concentration from 0.02% to 4% at room temperature could be detected within a response time of 10~30 s [70].

Organic Chemicals and Solute Detection

Due to the advantages of the technology and the detection platform, the use of PCFs for Raman spectroscopy is very common in chemical and biochemical sensing applications. Z. Xie et al. designed a broad spectral surface-enhanced Raman scattering (SERS) sensor by using a solid-core holey photonic crystal fiber with silver nanoparticles cluster. This SERS sensor was used to detect 4-Mercaptobenzoic acid. The strategy followed, contrary to previous works where high-pressure chemical deposition was employed, consisted of filling only 1 cm of fiber with Ag nanoparticles solution by capillary action. Compared to the signal obtained without Ag nanoparticles, higher intensities were obtained, as well as a good reproducibility of the results [71]. Yi Zhang et al. employed a hollow-core PCF to develop a fiber sensor based on SERS scattering. The hollow-core PCF was filled with liquid sample only in the core in order to maintain the bandgap and confine light in the hollow-filled region. More light energy interacting with the matter was achieved based on this method. The hollow-core PCF with a length of 10 cm was cut, and introduced to a 1-cm sample with silver nanoparticles (ranging from 40 to 60 nm in diameter) based on capillary action. Spectra of rhodamine 6 G, human insulin, and tryptophan were obtained by using the fiber as a Raman platform [72]. Not only Ag but also Au nanoparticles have been used to enhance the Raman signal. He Yan et al. proposed a SERS probe based on a hollow-core PCF in which Au nanoparticles were coated on the inner surface of air channels serving as a SERS substrate. A 5-cm long optical fiber was cut out and the Au nanoparticles (100–200 nm) colloidal was introduced (1–3 cm) by capillary action. After drying at 60 °C for 2 h, rhodamine B solution was introduced and dried. The signal of the analyte was verified after subtracting the background spectrum from the silica [73]. Xuan Yang et al. proposed a glucose biosensor based on a hollow-core PCF and Raman spectroscopy, as shown in Figure 4. An 8-cm long hollow-core PCF was used in the system, a liquid sample with D-glucose was introduced by capillary action, and then one end of the optical fiber was sealed. The peak signals from D-glucose measured in this system were in agreement with the direct analysis in a glass cuvette, while the intensity was 140-fold higher than the latter. Furthermore, they also prepared and measured solutions with different D-glucose concentrations, for which the LOD reached 1 mM [74].

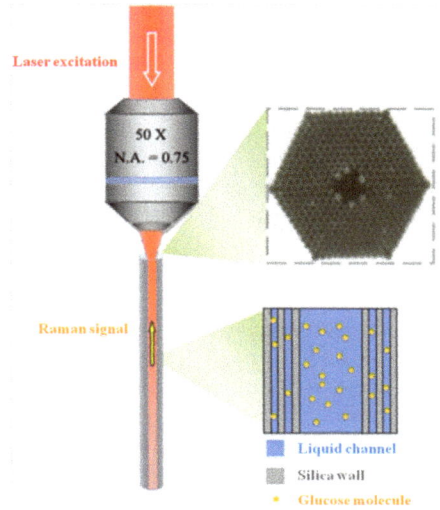

Figure 4. Schematic of a liquid-filled PCF probe for glucose detection proposed by Xuan Yang et al. Top right is cross-sectional view of the PCF probe. Bottom right depicts the air channels of the PCF filled with a glucose solution. Figure taken from Reference [74] with the permission of Springer Nature.

As for fluorescence measurements, Cordeiro et al. detected rhodamine in solid-core MOF and hollow-core MOFs by selectively filling MOFs [75]. The sample introduction and optical alignment are independent because they do not need to be in contact with the fiber tip, which is very advantageous for optical sensing in a laboratory. Stanislav O. Konorov et al. tested various fused silica and soft glass PCFs to detect a dye (thiacarbocyanine) [76]. The large-diameter cladding air channels were filled with a liquid sample based on microcapillarity. Diode-laser radiation was delivered to a sample through the central core of the fiber, and the fiber cladding collected the fluorescent response from the dye and guided it to a detector. Cordeiro C. M. et al. designed a new method to selectively fill solid-core PCFs and hollow-core PCFs with a liquid sample [77]. The air channels were collapsed in one fiber end firstl, and then pressure was applied from another fiber end when the arc of a fusion splicer softening the PCF near the point where the side-hole was generated. Lateral filling of rhodamine was achieved for both structures, and the substance was successfully detected. Williams et al. employed a 30-cm length of water filling a hollow-core MOF to measure fluorescein in the magnitude of the attomole and achieved LOD of 0.02 nM [78].

In addition, absorption-based detection has been successfully achieved with PCFs. Martelli et al. proposed using a pure silica solid-core PCF to detect porphyrin in an aqueous solution based on UV-Vis absorption. After filling a 22-cm long solid-core PCF by capillary action with an aqueous solution containing the sample, which took 3 h, detection was performed by using an excitation laser at 640 nm and a optical spectrum analyzer [79]. X. Yu et al. demonstrated an evanescent field absorption sensing technique in liquid solutions using a microstructured photonic crystal fiber. A defected-core PCF made of silica was used to detect the cobalt chloride (CoCl$_2$). This core consisted of a hole smaller than those from the cladding, and the evanescent field penetrated into the air channels, which enhanced the liquid absorption sensitivity. The air channels of the PCF were filled with the analyte solution by capillary action, following which concentrations of $CoCl_2$ from 0.01 M to 0.5 M were detected with a good linear response [80].

3.2.3. Biomedical Sensing inside Fluids

The air channels of MOFs serve as natural microfluidic channels, allowing MOFs to be potential biomedical sensors. Some researchers have demonstrated fiber-based biochemical sensors [81,82] through theoretical simulation, and explored the detection performance of biomolecules layer thickness. Md. Rabiul Hasan et al. numerically introduced surface plasmon resonance into the biosensor and further increased its sensitivity [83]. Zhengyong Li detected bovine serum albumin in experiments without specificity [84]. V. A. Popescu proposed a photonic fiber-based plasmonic sensor with a thin gold layer and 14 small air holes applied for the detection of human blood groups A, B, and O [85], which can also be applied for the determination of the hemoglobin concentration in normal human blood. Jian Sun et al. attempted to use goat antihuman IgG (antibody) for the specific detection of human IgG and yielded good results [86]. The sensitivity was found to reach around 0.1 nmol/L, and the required sample was less than 1 µL. Yi Zhang et al. employed a hollow-core PCF to develop a fiber sensor based on SERS scattering, and the spectra of human insulin and tryptophan were obtained [72].

Lei Wei's research group designed and fabricated a biosensor based on a side-channel photonic crystal fiber (SC-PCF), which is shown in Figure 5a,b [87]. One-sixth of the cladding air holes was left blank, and a larger air channel was formed on the side of the core. As shown in the inset of Figure 5b, the fabricated SC-PCF was spliced with a side-polished SMF to simultaneously enable lateral liquid channeling and light transmission in the fiber core. In this system, the electrostatic interaction-based modification method was used to immobilize the human cardiac troponin T (cTnT) antibody on the fiber core surface, as shown in Figure 6a. The method of biomass immobilization based on electrostatic interaction is relatively convenient. The various surface treatment layers consist of hydroxide ions generated from sodium hydroxide solution, a monolayer of poly(allylamine) (PAA), and a layer of the active cTnT antibody. In addition, bovine serum albumin (BSA) was added to prevent non-specific binding. In the final step, a group of cTnT protein solutions (concentration ranging from 1 pg/mL onwards) was pumped into the side-channel to characterize its sensing capability and the LOD. We can see from Figure 6a that with the completion of each step of treatment, the spectrum produced a corresponding shift. Figure 6b,c show the spectra and resonance wavelength shift under different concentrations of cTnT antigen solutions. The final experimental results show that the LOD of this biosensing system reached 1 ng/mL.

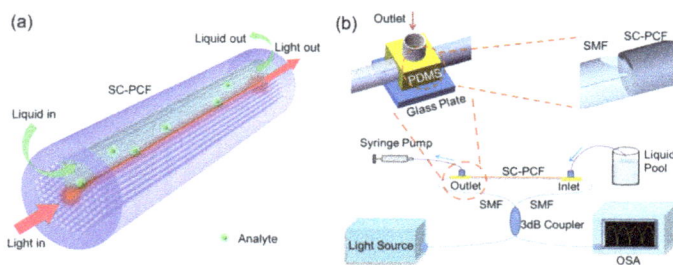

Figure 5. (**a**) Schematic diagram of liquid sensing with side-channel photonic crystal fiber (SC-PCF) (dimensions not to scale); (**b**) Details of the splicing point of the SC-PCF to side-polished single mode fibers (SMFs) and scheme of the absorption experiment; (Bottom) Scheme of the Sagnac interferometer. Figure taken from Reference [87] with the permission of the Royal Society of Chemistry.

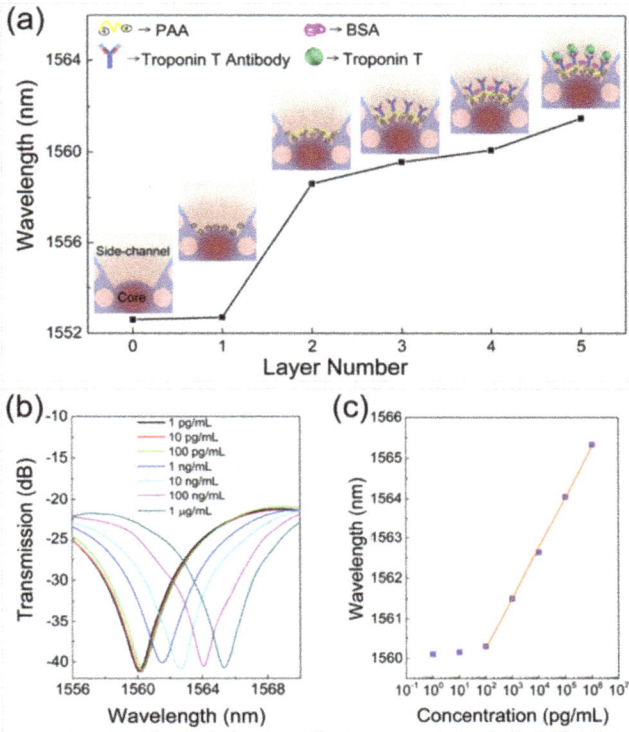

Figure 6. (a) Shift of the resonance wavelength monitored at various surface treatment steps (after rinsing thoroughly). The concentration of the human cardiac troponin T (cTnT) protein is 1 ng/mL. Insets are the illustrations of the binding profiles inside the side-channel. The red color region represents the mode profile in the fiber core and evanescent wave that extended to the side-channel. (b) Shift of the transmission spectra near the resonance wavelength of 1560 nm corresponding to the binding effect of different concentrations of the cTnT antigen. (c) Resonance wavelengths extracted from the spectra. The purple squares indicate experimental data and the orange straight line is linearly fitted to the experimental data. Figure taken from Reference [87] with the permission of the Royal Society of Chemistry.

J. B. Jensen et al. immobilized the antigen (streptavidin molecules) inside the air holes of a polymer MOF and realized the selective binding of fluorophore-labeled antibodies (α-streptavidin-Cy3) with antigen, while α-CRP-Cy3 molecules which could not be bound with streptavidin molecules were washed out [88]. If there were any antibodies in the analyte solution, the fluorophore could be detected so as to realize the selective detection of the target material. The corresponding specific detection principle is shown in Figure 7. In their work, the antigen was immobilized on the surface of PMMA directly, which is much more convenient than the treatment on the quartz fiber surface. This is because the streptavidin molecules can bind directly to the polymer surface while still being able to bind the antibody. Besides, the realization of the selective detection of α-streptavidin-Cy3 is based on selective binding between antigen and antibody, because of which the specific recognition is of high efficiency. The experimental results show that this biosensor had a detection limit of 80 nM with only a 27-μL sample volume.

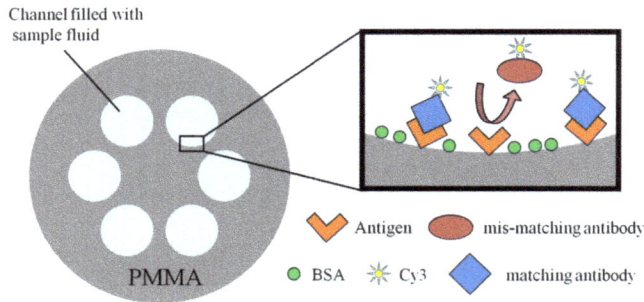

Figure 7. Schematic presentation of the capture processes utilized in the selective detection of the antibodies in Reference [88].

Xuan Yang et al. employed tip-coated multi-mode fibers (MMFs) and a hollow-core PCF with SERS, and realized the detection of protein lysozyme and cytochrome C as well as live bacterial cells of Shewanella oneidensis MR-1 in an aqueous solution. Two systems were proposed. A tip-coated MMF was charged positively with Ag nanoparticles and proteins mixed with other negatively charged Ag nanoparticles were introduced, which then formed a sandwich structure as shown in Figure 8. The LOD of this system was as low as 0.2 mg/mL, an order of magnitude lower than the direct measurement using a glass cuvette. Another approach involved utilizing the a hollow-core PCF with the cladding sealed and confined light in the core, filled with the bacteria under study, obtaining LOD of 106 cells/mL [89]. Dinish U.S. et al. proposed a SERS biosensor for cancer proteins in low sample volumes based on a hollow-core PCF. Epidermal growth factor receptors, a common biomarker for various cancers, were immobilized on the walls of the core and Au nanoparticles were attached for SERS measurements. After protein immobilization by capillary action, which has an order of magnitude higher strength than traditional materials (glass slides), the SERS experiments were carried out. By introducing an extremely low volume, an amount of approximately 100 pg of cancer protein could be detected, which improved the sensitivity compared to other methods previously reported in the literature [90].

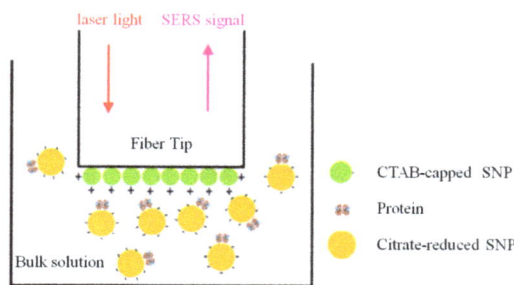

Figure 8. Scheme of the "sandwich" proposed by the authors for SERS sensing of proteins with an multi-mode fiber (MMF) (SERS: surface-enhanced Raman spectroscopy; CTAB: cetyltrimethylammonium bromide; SNP: silver nanoparticle). Reprinted (adapted) with permission from Reference [89]. Copyright (2011) American Chemical Society.

Another efficient method for the selective detection of a target material is based on molecularly imprinted polymer. A hybrid polymer containing target molecules was first produced on the surface of the optical fiber. Then the target molecules were separated from the polymer by hydrolysis, followed by rinsing with deionized water. As a consequence, some vacancies on the surface of the optical fiber were created, in which the target molecules can be embedded into exactly while other molecules are easily washed out. T. H. Nguyen et al. designed a fiber-optic biochemical sensor for the selective detection of cocaine based on molecularly imprinted polymer. The sensor detected an increasing fluorescence intensity in response to cocaine with an increasing concentration ranging from 0 to 500 µM in aqueous acetonitrile mixtures, and exhibited a good detection specificity compared with ketamine, ecgonine methyl ester, and other kinds of drugs [46]. Moreover, thanks to the stability of molecularly imprinted polymer, this kind of biosensor has good reproducibility. The biosensor reported in Reference [46] showed great selective detection performance even after one month.

From the biological protein concentration to the virus marker, it can be seen from the above investigation that the biosensor based on microstructured optical fiber has great potential. However, most of the current research remains in the theoretical simulation and preliminary experimental stages in the laboratory, and there is still a long way to go before mature practical application and productization. Lars Rindorf et al. have done a very good job in this area, as they presented the first incorporation of a microstructured optical fiber (MOF) into biochip applications, allowing the sample to flow continuously along the microstructure. Figure 9 shows a picture of this system and a scheme with the principal components in the sensing area of the system. A 16-mm long multi-core MOF of 125 µm outer diameter (OD), with a central air channels of 17 µm and a cladding composed of 312 channels of 2.3 µm, was integrated into a PMMA microchip. Theoretical calculations estimated that at the wavelength of 650 nm, around 6.5% of the field intensity, was in the air channels of the MOF. Techniques involving electrostatic coupling were employed to immobilize the DNA on the surface of the air channels of the MOF. After attaching, signals corresponding to the DNA were detected by comparing with the reference. Although this device had acceptable robustness, thermostatic control is required to improve its performance [91]. Yi Sun et al. took advantage of the exceptional heat-dissipation properties of PCFs and proposed to incorporate them into microcapillary electrophoresis chips. The PCF consists of a bundle of extremely narrow hollow channels, working as separation columns. A separation length of 7 cm was used in the PMMA microchip. After adding the dye (YOPRO-1) in the background electrolyte, DNA fragments could be detected by fluorescence measurements [92].

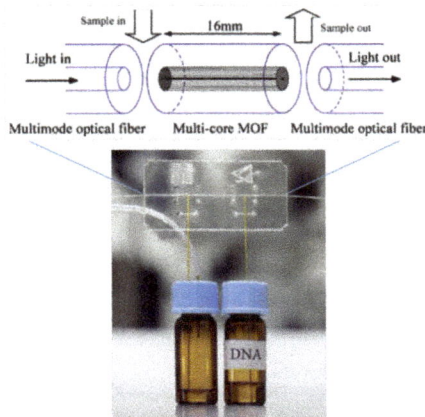

Figure 9. Scheme proposed by the authors where a lab-on-a-chip is integrated with an MOF, connected to two fibers to guide light. Fluid samples are injected through silica capillary tubes. Above, a scheme of the sensing area of the system is enlarged. Figure taken from Reference [91] with the permission of Springer.

3.3. Other Applications

The magnetic nanoparticles solutions sensitive to an external magnetic field were infiltrated into PCFs. Harneet V. Thakur et al. reported a magnetic field sensor based on a polarization-maintaining PCF that infiltrated small amount of Fe_3O_4 magnetic optofluid/nanofluid in cladding holes, and a higher sensitivity of 242 pm/mT was realized [93]. Sully M. M. Quintero et al. proposed a magnetic field sensor comprised of a high birefringence photonic crystal fiber coated with a Terfenol-D/Epoxy composite layer, and the sensitivity of the developed sensor with magnetic fields was measured to be 6 pm/mT [94].

If the air channels of a PCF are filled with a substance that exhibits a strong electro-optical effect, particularly in liquid crystal that can effectively modify those optical properties through an electric field, the optofluidic PCF can also be used as an effective sensor for the electric field [95,96].

Super-continuous generation in the hollow-core PCF should also be mentioned, since it is one of the few PCF applications that has led to commercial products. Prathamesh S. Donvalkar et al. demonstrated high optical depths (50 ± 5) in Rubidium-filled hollow-core photonic band-gap fibers, which represented a 1000-fold improvement over operation times previously reported, in addition to lasting for hours. They also studied the vapor generation mechanism using continuous-wave and pulsed light sources and found that the mechanism that generates the Rubidium atoms is actually due to thermal evaporation [97].

4. Conclusions

In this paper, we reviewed the development and applications of optofluidics implemented based on MOFs, which have been demonstrated to be a promising platform with miniature size. The types of MOFs, especially those with hollow cores, are presented to emphasize the flexibility of designing air-hole structures for optofluidics. There are several main approaches to the fabrication of the inlet/outlet summarized in this study, which allow fluidics to enter the air holes of MOFs. In practical application, the approaches of drilling holes or using C-shaped fibers are more robust and reliable to achieve optofluidics. These approaches render it possible to make the entire system more compact compared with those using the butt-coupling method or other connection tools. Moreover, this technique is suitable for various kinds of MOFs, which then provides more flexibilities of configuring the optofluidic platform employing MOFs.

With a suitable design, particularly for biochemical applications, the MOF-based optofluidic platform can allow highly sensitive devices to sense fluidic flow rate, refractive index, and pH value, as well as conduct material analyses. Furthermore, it is also a potential tool to demonstrate physical phenomena with a small volume of materials. As a widely used technology, the fluorescence excited by specific substance filled in MOFs enables spectroscopic detection for biosensing. Among these, photochemistry is the most promising, especially when the system size and efficiency are critical concerns. Materials analysis could be one of the potential applications, as reviewed in this paper. The lab-in-fiber technology in MOFs provides the possibility of conducting chemical reactions only in the air channels of MOFs, which only requires a small volume.

For future prospects, the coupling of light and liquids into MOFs still raises research issues that attract attention. There are flexible air holes in MOFs, which means that these holes can be selectively filled with various liquids in order to conduct analyses. However, higher precision alignment and a better design of the inlet/outlet are required. Like the lab-on-chip technology, the microfluidic channels of an MOF can be designed and fabricated in other forms. In terms of good interaction between matter and light, the hollow-core MOFs and index-guiding MOFs with thin cores are better choices; however, their fabrication requires strict conditions and could be one future research topic aiming at the applications of optofluidics.

To conclude, optofluidics in MOFs open a whole new sector of useful applications for material characterization. The capability of propagating fluids and light simultaneously in MOFs makes them an excellent and encouraging platform for optofluidics.

Acknowledgments: L.S. and J.H. would like to acknowledge the funding support from the National Natural Science Foundation of China (61475128). Z.L., D.G. and H.-Y.T. would like to acknowledge the funding support from the Hong Kong Polytechnic University under 1-ZVGB.

Author Contributions: L.S., Z.L., J.H., D.G. co-wrote the manuscript, H.-Y.T. supervised the work. All of the authors discussed the work and revised the paper.

Conflicts of Interest: The authors declare no conflict of interest.

References

1. Whitesides, G.M. The origins and the future of microfluidics. *Nature* **2006**, *442*, 368. [CrossRef] [PubMed]
2. Bykov, D.S.; Schmidt, O.A. Flying particle sensors in hollow-core photonic crystal fibre. *Nat. Photonics* **2015**, *9*, 461. [CrossRef]
3. Unterkofler, S.; Mcquitty, R.J. Microfluidic integration of photonic crystal fibers for online photochemical reaction analysis. *Opt. Lett.* **2012**, *37*, 1952–1954. [CrossRef] [PubMed]
4. Warrensmith, S.C.; Ebendorffheidepriem, H. Exposed-core microstructured optical fibers for real-time fluorescence sensing. *Opt. Express* **2009**, *17*, 18533–18542. [CrossRef] [PubMed]
5. Yuan, Y.; Guo, T. Electrochemical Surface Plasmon Resonance Fiber-Optic Sensor: In Situ Detection of Electroactive Biofilms. *Anal. Chem.* **2016**, *88*, 7609–7616. [CrossRef] [PubMed]
6. Liu, Z.; Zhang, A.P. Microfluidic device integrated with FBG in Co^{2+}-doped fiber to measure flow rate with nL/s sensitivity. In Proceedings of the 23rd International Conference on Optical Fibre Sensors, Santander, Spain, 2–6 June 2014; SPIE: Bellingham, WA, USA, 2014; Volume 9157, p. 91573I.
7. Lien, V.; Vollmer, F. Microfluidic flow rate detection based on integrated optical fiber cantilever. *Lab Chip* **2007**, *7*, 1352. [CrossRef] [PubMed]
8. Minzioni, P.; Osellame, R. Roadmap for optofluidics. *J. Opt.* **2017**, *19*, 093003. [CrossRef]
9. Haeberle, S.; Zengerle, R. Microfluidic platforms for lab-on-a-chip applications. *Lab Chip* **2007**, *7*, 1094–1110. [CrossRef] [PubMed]
10. Cusano, A.; Ricciardi, A. *Lab-on-Fiber Technology*; Springer: Basel Switzerland, 2015; Volume 56.
11. Consales, M.; Ricciardi, A. Lab-on-Fiber Technology: Toward Multifunctional Optical Nanoprobes. *ACS Nano* **2012**, *6*, 3163. [CrossRef] [PubMed]
12. Zhang, Q.; Hao, P. High-visibility in-line fiber-optic optofluidic Fabry–Pérot cavity. *Appl. Phys. Lett.* **2017**, *111*, 191102. [CrossRef]
13. Zhang, C.L.; Gong, Y. Microbubble-Based Fiber Optofluidic Interferometer for Sensing. *J. Lightwave Technol.* **2017**, *35*, 2514–2519. [CrossRef]
14. Domachuk, P.; Littler, I.C.M. Compact resonant integrated microfluidic refractometer. *Appl. Phys. Lett.* **2006**, *88*, 330A. [CrossRef]
15. Monat, C.; Domachuk, P. Integrated optofluidics: A new river of light. *Nat. Photonics* **2007**, *1*, 106–114. [CrossRef]
16. Leite, I.T.; Turtaev, S. Three-dimensional holographic optical manipulation through a high-numerical-aperture soft-glass multimode fibre. *Nat. Photonics* **2018**, *12*, 33. [CrossRef]
17. Cubillas, A.M.; Unterkofler, S. Photonic crystal fibres for chemical sensing and photochemistry. *Chem. Soc. Rev.* **2013**, *42*, 8629–8648. [CrossRef] [PubMed]
18. Cubillas, A.M.; Schmidt, M. Ultra-Low Concentration Monitoring of Catalytic Reactions in Photonic Crystal Fiber. *Chem. A Eur. J.* **2012**, *18*, 1586–1590. [CrossRef] [PubMed]
19. Jensen, J.B.; Pedersen, L.H. Photonic crystal fiber based evanescent-wave sensor for detection of biomolecules in aqueous solutions. *Opt. Lett.* **2004**, *29*, 1974–1976. [CrossRef] [PubMed]
20. Ertman, S.; Lesiak, P. Optofluidic photonic crystal fiber-based sensors. *J. Lightwave Technol.* **2017**, *35*, 3399–3405. [CrossRef]
21. Knight, J.C.; Birks, T.A. All-silica single-mode optical fiber with photonic crystal cladding. *Opt. Lett.* **1996**, *21*, 1547. [CrossRef] [PubMed]
22. Wu, C.; Tam, H.Y. Fabrication, Characterization, and Sensing Applications of a High-Birefringence Suspended-Core Fiber. *J. Lightwave Technol.* **2014**, *32*, 2113–2122.
23. Monro, T.M.; Ebendorff-Heidepriem, H. Sensing in suspended-core optical fibers. *Opt. Fiber Technol.* **2010**, *16*, 343–356. [CrossRef]

24. Russell, P.S.J.; Knight, J.C. Photonic crystal fibres. *Nature* **2003**, *424*, 847–851.

25. Litchinitser, N.M.; Abeeluck, A.K. Antiresonant reflecting photonic crystal optical waveguides. *Opt. Lett.* **2002**, *27*, 1592. [CrossRef] [PubMed]

26. Wu, D.K.; Kuhlmey, B.T. Ultrasensitive photonic crystal fiber refractive index sensor. *Opt. Lett.* **2009**, *34*, 322–324. [CrossRef] [PubMed]

27. Chemnitz, M.; Gebhardt, M. Hybrid soliton dynamics in liquid-core fibres. *Nat. Commun.* **2017**, *8*, 42. [CrossRef] [PubMed]

28. Wu, C.; Tse, M.L.V. In-line microfluidic refractometer based on C-shaped fiber assisted photonic crystal fiber Sagnac interferometer. *Opt. Lett.* **2013**, *38*, 3283–3286. [CrossRef] [PubMed]

29. Wang, Y.; Liu, S. Selective-Fluid-Filling Technique of Microstructured Optical Fibers. *J. Lightwave Technol.* **2010**, *28*, 3193–3196.

30. Gao, R.; Lu, D. Simultaneous measurement of refractive index and flow rate using graphene-coated optofluidic anti-resonant reflecting guidance. *Opt. Express* **2017**, *25*, 28731. [CrossRef]

31. Yang, X.; Guo, X. Lab-on-fiber electrophoretic trace mixture separating and detecting an optofluidic device based on a microstructured optical fiber. *Opt. Lett.* **2016**, *41*, 1873. [CrossRef] [PubMed]

32. Tian, F.; Min, J. Lab-on-fiber optofluidic platform for in situ monitoring of drug release from therapeutic eluting polyelectrolyte multilayers. *Opt. Express* **2015**, *23*, 20132–20142. [CrossRef] [PubMed]

33. Jelger, P.; Laurell, F. An In-Fibre Microcavity. In Proceedings of the 2007 Conference on Lasers and Electro-Optics, Baltimore, MD, USA, 6–11 May 2007.

34. Sudirman, A.; Margulis, W. All-Fiber Optofluidic Component to Combine Light and Fluid. *IEEE Photonics Technol. Lett.* **2014**, *26*, 1031–1033. [CrossRef]

35. Tian, F.; Kanka, J. Photonic crystal fiber for layer-by-layer assembly and measurements of polyelectrolyte thin films. *Opt. Lett.* **2012**, *37*, 4299–4301. [CrossRef] [PubMed]

36. Nissen, M.; Doherty, B. UV Absorption Spectroscopy in Water-Filled Antiresonant Hollow Core Fibers for Pharmaceutical Detection. *Sensors* **2018**, *18*, 478. [CrossRef] [PubMed]

37. Tiwari, V.S.; Kalluru, R.R. Fiber optic Raman sensor to monitor the concentration ratio of nitrogen and oxygen in a cryogenic mixture. *Appl. Opt.* **2007**, *46*, 3345–3351. [CrossRef] [PubMed]

38. Motz, J.T.; Hunter, M. Optical Fiber Probe for Biomedical Raman Spectroscopy. *Appl. Opt.* **2004**, *43*, 542. [CrossRef] [PubMed]

39. Rouvie, A.; Roosen, G. Stimulated Raman scattering in an ethanol core microstructured optical fiber. *Opt. Express* **2005**, *13*, 4786–4791.

40. Cox, F.M.; Argyros, A. Surface enhanced Raman scattering in a hollow core microstructured optical fiber. *Opt. Express* **2007**, *15*, 13675–13681. [CrossRef] [PubMed]

41. Amezcua-Correa, A.; Yang, J. Surface-Enhanced Raman Scattering Using Microstructured Optical Fiber Substrates. *Adv. Funct. Mater.* **2010**, *17*, 2024–2030. [CrossRef]

42. Yan, H.; Liu, J. Novel index-guided photonic crystal fiber surface-enhanced Raman scattering probe. *Opt. Express* **2008**, *16*, 8300–8305. [CrossRef] [PubMed]

43. Wang, G.; Liu, J. Gas Raman sensing with multi-opened-up suspended core fiber. *Appl. Opt.* **2011**, *50*, 6026. [CrossRef] [PubMed]

44. Afshar, V.S.; Ruan, Y. Enhanced fluorescence sensing using microstructured optical fibers: a comparison of forward and backward collection modes. *Opt. Lett.* **2008**, *33*, 1473–1475. [CrossRef]

45. Warren-Smith, S.C.; Afshar, S. Theoretical study of liquid-immersed exposed-core microstructured optical fibers for sensing. *Opt. Express* **2008**, *16*, 9034. [CrossRef] [PubMed]

46. Nguyen, T.H.; Hardwick, S.A. Intrinsic Fluorescence-Based Optical Fiber Sensor for Cocaine Using a Molecularly Imprinted Polymer as the Recognition Element. *IEEE Sens. J.* **2011**, *12*, 255–260. [CrossRef]

47. Chu, F.; Tsiminis, G. Explosives detection by fluorescence quenching of conjugated polymers in suspended core optical fibers. *Sens. Actuators B Chem.* **2014**, *199*, 22–26. [CrossRef]

48. Yuan, L.; Yang, J. A Compact Fiber-Optic Flow Velocity Sensor Based on a Twin-Core Fiber Michelson Interferometer. *IEEE Sens. J.* **2008**, *8*, 1114–1117. [CrossRef]

49. Lu, P.; Chen, Q. Fiber Bragg grating sensor for simultaneous measurement of flow rate and direction. *Meas. Sci. Technol.* **2008**, *19*, 1169–1175. [CrossRef]

50. Li, Y.; Yan, G. Microfluidic flowmeter based on micro "hot-wire" sandwiched Fabry-Perot interferometer. *Opt. Express* **2015**, *23*, 9483. [CrossRef] [PubMed]

51. Christodoulides, P.; Florides, G.A. 1–Microfluidics flow and heat transfer in microstructured fibers of circular and elliptical geometry. In *Optofluidics Sensors & Actuators in Microstructured Optical Fibers*; Elsevier: Amsterdam, The Netherlands, 2015; pp. 3–27.

52. Wang, G.; Liu, J. Fluidic sensor based on the side-opened and suspended dual-core fiber. *Appl. Opt.* **2012**, *51*, 3096. [CrossRef] [PubMed]

53. Wang, G.; Wang, C. Side-Opened Suspended Core Fiber-Based Surface Plasmon Resonance Sensor. *IEEE Sens. J.* **2015**, *15*, 4086–4092. [CrossRef]

54. Hassani, A.; Kabashin, A. Photonic bandgap fiber-based Surface Plasmon Resonance sensors. *Opt. Express* **2007**, *15*, 11413.

55. Yuan, W.; Town, G.E. Ultrasensitive refractive index sensor based on twin-core photonic bandgap fibers. In Proceedings of the 20th International Conference on Optical Fibre Sensors, Edinburgh, UK, 5–9 October 2009; Volume 7503, p. 75035A.

56. Yu, X.; Shum, P. Highly Sensitive Photonic Crystal Fiber-Based Refractive Index Sensing Using Mechanical Long-Period Grating. *IEEE Photonics Technol. Lett.* **2008**, *20*, 1688–1690. [CrossRef]

57. Lin, W.; Zhang, H. Microstructured optical fiber for multichannel sensing based on Fano resonance of the whispering gallery modes. *Opt. Express* **2017**, *25*, 994. [CrossRef] [PubMed]

58. Luan, N.; Ding, C. A Refractive Index and Temperature Sensor Based on Surface Plasmon Resonance in an Exposed-Core Microstructured Optical Fiber. *IEEE Photonics J.* **2016**, *8*, 1–8. [CrossRef]

59. Cox, F.M.; Lwin, R. Opening up optical fibres. *Opt. Express* **2007**, *15*, 11843–11848. [CrossRef] [PubMed]

60. Yang, X.H.; Wang, L.L. Fluorescence pH probe based on microstructured polymer optical fiber. *Opt. Express* **2007**, *15*, 16478–16483. [CrossRef] [PubMed]

61. Hoo, Y.L.; Jin, W. Evanescent-wave gas sensing using microstructure fiber. *Opt. Eng.* **2002**, *41*, 8–9. [CrossRef]

62. Fini, J.M. Microstructure fibres for optical sensing in gases and liquids. *Meas. Sci. Technol.* **2004**, *15*, 1120–1128. [CrossRef]

63. Li, S.G.; Liu, S.Y. Study of the sensitivity of gas sensing by use of index-guiding photonic crystal fibers. *Appl. Opt.* **2007**, *46*, 5183. [CrossRef] [PubMed]

64. Ritari, T.; Tuominen, J. Gas sensing using air-guiding photonic bandgap fibers. *Opt. Express* **2004**, *12*, 4080–4087. [CrossRef] [PubMed]

65. Hoo, Y.L.; Jin, W. Gas diffusion measurement using hollow-core photonic bandgap fiber. *Sens. Actuators B Chem.* **2005**, *105*, 183–186. [CrossRef]

66. Pawłat, J.; Sugiyama, T. PBG Fibers for Gas Concentration Measurement. *Plasma Process. Polym.* **2007**, *4*, 743–752. [CrossRef]

67. Gayraud, N.; Kornaszewski, Ł.W. Mid-infrared gas sensing using a photonic bandgap fiber. *Appl. Opt.* **2008**, *47*, 1269–1277. [CrossRef] [PubMed]

68. Hoo, Y.L.; Liu, S. Fast Response Microstructured Optical Fiber Methane Sensor with Multiple Side-Openings. *IEEE Photonics Technol. Lett.* **2010**, *22*, 296–298. [CrossRef]

69. Kassani, S.H.; Khazaeinezhad, R. Suspended Ring-Core Photonic Crystal Fiber Gas Sensor with High Sensitivity and Fast Response. *IEEE Photonics J.* **2015**, *7*, 1–9. [CrossRef]

70. Zhou, X.; Dai, Y. Microstructured FBG hydrogen sensor based on Pt-loaded WO3. *Opt. Express* **2017**, *25*, 8777. [CrossRef] [PubMed]

71. Xie, Z.; Lu, Y. Broad spectral photonic crystal fiber surface enhanced Raman scattering probe. *Appl. Phys. B* **2009**, *95*, 751–755. [CrossRef]

72. Zhang, Y.; Shi, C. Liquid core photonic crystal fiber sensor based on surface enhanced Raman scattering. *Appl. Phys. Lett.* **2007**, *90*, 193504. [CrossRef]

73. Yan, H.; Gu, C. Hollow core photonic crystal fiber surface-enhanced Raman probe. *Appl. Phys. Lett.* **2006**, *89*, 1102. [CrossRef]

74. Yang, X.; Zhang, A.Y. Direct molecule-specific glucose detection by Raman spectroscopy based on photonic crystal fiber. *Anal. Bioanal. Chem.* **2012**, *402*, 687–691. [CrossRef] [PubMed]

75. Cordeiro, C.M.B.; Franco, M.A.R. Microstructured-core optical fibre for evanescent sensing applications. *Opt. Express* **2006**, *14*, 13056–13066. [CrossRef] [PubMed]

76. Zheltikov, A.M.; Scalora, M. Photonic-crystal fiber as a multifunctional optical sensor and sample collector. *Opt. Express* **2005**, *13*, 3454–3459.

77. Cordeiro, C.M.; Dos Santos, E.M. Lateral access to the holes of photonic crystal fibers—Selective filling and sensing applications. *Opt. Express* **2006**, *14*, 8403–8412. [CrossRef] [PubMed]

78. Williams, G.; Euser, T.G. Spectrofluorimetry with attomole sensitivity in photonic crystal fibres. *Methods Appl. Fluoresc.* **2013**, *1*, 015003. [CrossRef] [PubMed]

79. Martelli, C.; Canning, J. Water-soluble porphyrin detection in a pure-silica photonic crystal fiber. *Opt. Lett.* **2006**, *31*, 2100–2102. [CrossRef] [PubMed]

80. Yu, X.; Sun, Y. Evanescent Field Absorption Sensor Using a Pure-Silica Defected-Core Photonic Crystal Fiber. *IEEE Photonics Technol. Lett.* **2008**, *20*, 336–338. [CrossRef]

81. Markos, C.; Yuan, W. Label-free biosensing with high sensitivity in dual-core microstructured polymer optical fibers. *Opt. Express* **2011**, *19*, 7790. [CrossRef] [PubMed]

82. Rindorf, L.; Jensen, J.B. Photonic crystal fiber long-period gratings for biochemical sensing. *Opt. Express* **2006**, *14*, 8224. [CrossRef] [PubMed]

83. Hasan, M.R.; Akter, S. Spiral Photonic Crystal Fiber Based Dual- Polarized Surface Plasmon Resonance Biosensor. *IEEE Sens. J.* **2017**, *18*, 133–140. [CrossRef]

84. Li, Z.; Liao, C.; Chen, D.; Song, J.; Jin, W.; Peng, G.D.; Zhu, F.; Wang, Y.; He, J.; Wang, Y. Label-free detection of bovine serum albumin based on an in-fiber Mach-Zehnder interferometric biosensor. *Opt. Express* **2017**, *25*, 17105. [CrossRef] [PubMed]

85. Popescu, V.A. Application of a Plasmonic Biosensor for Detection of Human Blood Groups. *Plasmonics* **2017**, *12*, 1733–1739. [CrossRef]

86. Sun, J.; Chan, C.C. High-resolution photonic bandgap fiber-based biochemical sensor. *J. Biomed. Opt.* **2007**, *12*, 044022. [CrossRef] [PubMed]

87. Zhang, N.; Li, K. Ultra-sensitive chemical and biological analysis via specialty fibers with built-in microstructured optofluidic channels. *Lab Chip* **2018**. [CrossRef] [PubMed]

88. Jensen, J.; Hoiby, P. Selective detection of antibodies in microstructured polymer optical fibers. *Opt. Express* **2005**, *13*, 5883–5889. [CrossRef] [PubMed]

89. Yang, X.; Gu, C. Highly Sensitive Detection of Proteins and Bacteria in Aqueous Solution Using Surface-Enhanced Raman Scattering and Optical Fibers. *Anal. Chem.* **2011**, *83*, 5888–5894. [CrossRef] [PubMed]

90. Dinish, U.S.; Fu, C.Y.; Soh, K.S.; Ramaswamy, B.; Kumar, A.; Olivo, M. Highly sensitive SERS detection of cancer proteins in low sample volume using hollow core photonic crystal fiber. *Biosens. Bioelectron.* **2012**, *33*, 293.

91. Rindorf, L.; Høiby, P.E. Towards biochips using microstructured optical fiber sensors. *Anal. Bioanal. Chem.* **2006**, *385*, 1370–1375. [CrossRef] [PubMed]

92. Sun, Y.; Nguyen, N.T. Enhanced electrophoretic DNA separation in photonic crystal fiber. *Anal. Bioanal. Chem.* **2009**, *394*, 1707–1710. [CrossRef] [PubMed]

93. Thakur, H.V.; Nalawade, S.M. Photonic crystal fiber injected with Fe_3O_4 nanofluid for magnetic field detection. *Appl. Phys. Lett.* **2011**, *99*, 161101. [CrossRef]

94. Quintero, S.M.M.; Martelli, C. Photonic crystal fiber sensor for magnetic field detection. In Proceedings of the Fourth European Workshop on Optical Fibre Sensors, Porto, Portugal, 8–10 September 2010; SPIE: Bellingham, WA, USA, 2010; Volume 7653.

95. Bjarklev, A.; Xianyu, H. Frequency tunability of solid-core photonic crystal fibers filled with nanoparticle-doped liquid crystals. *Opt. Express* **2009**, *17*, 3754–3764.

96. Budaszewski, D.; Srivastava, A.K. Photo-aligned photonic ferroelectric liquid crystal fibers. *J. Soc. Inf. Disp.* **2015**, *23*, 196–201. [CrossRef]

97. Donvalkar, P.S.; Ramelow, S. Continuous generation of rubidium vapor in hollow-core photonic bandgap fibers. *Opt. Lett.* **2015**, *40*, 5379–5382. [CrossRef] [PubMed]

micromachines

MDPI

Review

Optofluid-Based Reflective Displays

Mingliang Jin [1,2], **Shitao Shen** [1,2], **Zichuan Yi** [3], **Guofu Zhou** [1,2] **and Lingling Shui** [1,2,3,*]

[1] National Center for International Research on Green Optoelectronics, South China Normal University, Guangzhou 510006, China; jinml@scnu.edu.cn (M.J.); shenshitao@m.scnu.edu.cn (S.S.); guofu.zhou@m.scnu.edu.cn (G.Z.)

[2] Guangdong Provincial Key Laboratory of Optical Information Materials and Technology and Institute of Electronic Paper Displays, South China Academy of Advanced Optoelectronics, South China Normal University, Guangzhou 510006, China

[3] Zhongshan Institute, University of Electronic Science and Technology of China, Zhongshan 528402, China; yizichuan@zsc.edu.cn

* Correspondence: Shuill@m.scnu.edu.cn; Tel.: +86-20-3931-4813

Received: 6 February 2018; Accepted: 27 March 2018; Published: 1 April 2018

Abstract: Displays can present information like text, images, or videos in a different color (visible light) by activating the materials in pixels. In a display device, pixels are typically of micrometer size and filled with displaying materials that are aligned and controlled by a display driver integrated circuit. Typical reflective displays can show designed information by manipulating ambient light via the microfluidic behavior in pixels driven by electrophoresis, electrowetting, or electromechanical forces. In this review, we describe the basic working principles and device structures of three reflective displays of electrophoresis display (EPD), electrowetting display (EWD), and interferometric modulator display (IMOD). The optofluidic behavior and controlling factors relating to the display performance are summarized.

Keywords: reflective display; electrowetting; electrophoresis; interferometric modulator; MEMS

1. Introduction

Displays show information on demand by manipulating visible light via changing the colors of one material or moving different colored materials at the microscale pixels. Depending on the interaction between light and materials, visible light can be manifested by different materials via reflection, transmission, or emission. Liquid crystal display (LCD), organic light emission display (OLED), and electrophoretic display (EPD) are the representatives of transmissive display, emissive display, and reflective display, respectively. In reflective displays, visible light with a wavelength in the range of about 400–700 nm could be either reflected, scattered, or absorbed by the displaying materials in pixels. Each pixel with a size in the range of nano- to micrometers is controlled by a corresponding electrical backplane, showing either different colors or grayscale (degree of black-and-white) [1–3].

Reflective display is also named "electronic paper," which possesses both the speed of "electronics" and the reading comfort of "paper." Thus, it has been given a lot of attention over the last half-century. The commercially available Kindle e-Paper and electronic batch are successful examples of EPD [4–6].

A schematic drawing of a reflective display device is shown in Figure 1. It consists of basic components (from bottom to top) of bottom substrate → driving electrical layer (thin film transistors or electrodes) → displaying material layer (in pixels or not) → top electrode(s) → top (cover) substrate including protective and optical films.

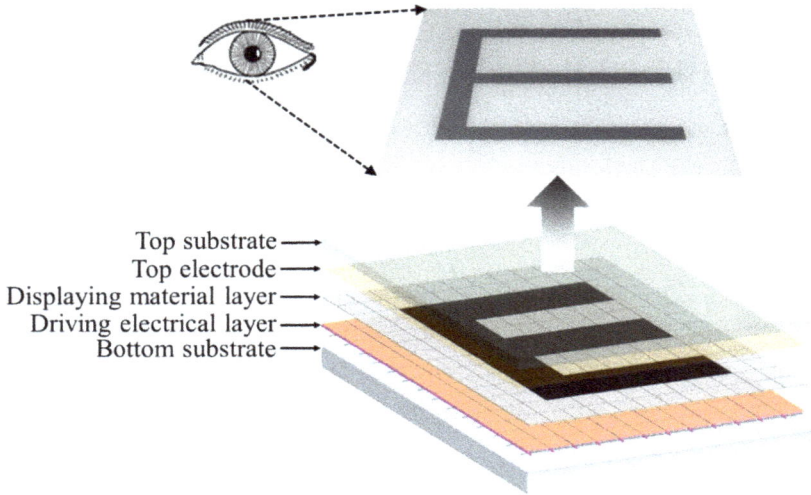

Figure 1. Schematic of a reflective display device.

Pixel size in display device is typically in the micrometer range at the horizontal x- and y-axis, and nanometer to micrometer in the vertical z-axis. Thus, human eyes cannot directly distinguish pixels from each other. However, once an area of pixels is actuated to demonstrate the same color, obvious light/color differences are visible by eyes. Each pixel can be driven by a display backplane individually to achieve a complex information display. For instance, a black "*E*" could be shown on a white background by only driving the pixels in the "*E*" area to black color. From this point of view, each pixel can be regarded as an optical switch. With reversible control of the displaying materials, the numbers and positions of the pixels will determine the content of displayed information. Therefore, the switching (open and closed) speed and degree determine the quality of a display device in speed and contrast ratio, respectively.

In this work, we define "optofluid-based reflective display" as a reflective display using interactions between light (visible light) and fluids (e.g., dispersion in microcapsule for EPD, dual-fluid in micropixel for EWD, and air in microcavity for IMOD). The electrochemical reaction (for electrochromic display) and molecular re-arrangement (for liquid crystal display) driven by the electric field are excluded. In an electrophoretic display (EPD), black and white particles suspended in an insulating liquid medium are driven by electrophoresis to move up and down to show black, white, or gray [7]. In an electrowetting display (EWD), dye dissolved organic or aqueous solutions are driven to move by electrowetting to display the color of the liquid solution or the background substrate [2,8]. In an interferometric modulator display (IMOD), the incident light is reflected by the metal bottom surface and the stacked film, showing the color of the constructively interfered wavelength by controlling the height of the air cavity via electromechanical force [3].

2. Electrophoretic Display

Electrophoretic display was first presented in 1973 [5] by placing two bi-chromo-color pigments in glass cavities and controlling the movement of particles to switch between two colors. Electrophoretic displays have been the subject of intense research for many years because of their wide marketing potential [7,9,10]. Currently, EPDs have been widely used in the fields of electronic books, electronic labels, and smart watches [4,11].

The major breakthrough of EPDs was made in 1998 by Jacobson et al. [7], who employed electrophoretic materials and device design to overcome the critical shortcomings of EPDs at that time. The mature and most successful EPD mode is named microcapsule-based EPD, in which the dispersion of black and white nanoparticles is encapsulated in microscale capsules and driven by pixelated electric backplane. The microcapsule-based EPD is shown in Figure 2. The black and white particles encapsulated in the microcapsules are positively and negatively charged, respectively. The microcapsules are sandwiched between the top transparent electrode and bottom pixel electrodes. The pixel electrodes can provide alternative voltages to actuate white and black particles to move in the microcapsules along the direction of electric field. In order to achieve accurate gray tones, the applied voltage on a common electrode can be adjusted. Therefore, display performance like grayscale or contrast ration is closely related to the applied voltage.

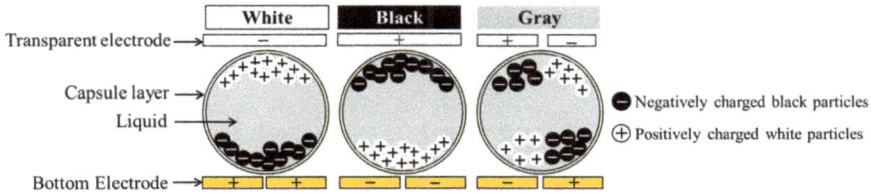

Figure 2. Working principle of the microcapsule-based electrophoretic display.

In an EPD pixel (microcapsule), typically, five chemical materials exist: pixel wall (or microcapsule shell), insulating liquid, colored pigments (particles), charge control agent, and stabilizer. Electrophoresis causes the movement of charged particles through the stationary insulating liquid. The conventional Helmholtz–Smoluchowski equation is applied in EPDs [12]:

$$U = \frac{\varepsilon \zeta_{EP} E_x}{\mu} \tag{1}$$

where U is the electrophoretic velocity of the particle, ε is the dielectric constant of the insulating liquid, ζ_{EP} is the zeta potential of the particle, E_x is the applied electrical field, and μ is the mobility of the particle. The electrophoretic zeta potential (ζ_{EP}) is a property of the charged particle. Except for electrophoretically driven particle moving, a stabilizer is used to keep positively charged white particles and negatively charged black particles separate from each. Therefore, the electrophoretic particles are typically core-shell structure with white or black cores encapsulated by polymeric shells. Moreover, by tuning the coating polymeric structure, the density difference between the insulating liquid, black particles, and white particles could be further manipulated to form a stable product for long-term usage.

The displaying material in EPDs is a colloidal suspension that is similar to painting and printing inks; it thus allows for comfortable readability in natural environmental conditions with a similar contrast ratio to newspapers and a reflectivity of ~40%. For the same reason, the displaying materials of EPDs are called "E-ink." Moreover, the microscale encapsulation minimizes the gravity effect; thus, the force balance without external voltage could be achieved. Therefore, EPDs possess the advantage of "bi-stability" for low energy consumption. Generally, an e-book reader can work for two weeks without recharging, and an image can be maintained for years without driving actuation. Furthermore, the spherical shape of the microcapsules makes the view angle wide without an obvious difference. However, the display speed of ~10 fps, with just several options of ink colors, has limited its applications. On the other hand, flexible displays could be easily fabricated using E-ink materials through wet-printing technologies. This makes it possible for wider applications in wearable electronic devices and electronic skins.

3. Electrowetting Display

The electrowetting principle, as applied in display devices, was first demonstrated in 1981 by Beni et al. [13–16]. The working principle of EWD is based on electrowetting driving microdroplets to move laterally or vertically. A practical and well-functional electrowetting display device based on colored oil driven by conductive liquid was successfully demonstrated in 2003 by Hayes and Feenstra at Philips Research Lab [2]. EWDs attracted much attention thereafter. In 2009, a brilliant EWD was presented by Heikenfeld et al., in which pigment dispersions were driven to move vertically between two layers in a multilayer structure [8].

A schematic structure of the dual microfluidic EWD is shown in Figure 3. It consists of a bottom substrate covered with pixel electrodes, hydrophobic insulating layer, hydrophilic pixel walls, colored oil (displaying material), conductive liquid, and top electrodes attached on the top substrate. At the stage of 0 voltage, the colored oil forms a continuous film in a pixel between the hydrophobic insulator (connected with the bottom electrode) and the conductive liquid (connected with the top electrode) due to the hydrophobic interaction between the oil and hydrophobic coating. At this stage, the pixel shows the color of the oil. When voltage V is applied between the top and bottom electrodes, the energetically stable state is destroyed since an electrostatic force is added. Water (a conductive liquid) moves towards the bottom surface driven by electrowetting, breaking and pushing the oil film aside to pixel walls and corners. At this stage, the pixel shows the color of the bottom substrate when viewed from the top. In this way, the optical properties of the stack are tuned between an OFF-state color (e.g., black oil) and an ON-state color (e.g., white substrate). In general, specific grayscale can be achieved by the oil film coverage ratio over a pixel, being controlled by the applied voltage across the top and bottom electrodes. Overall, a simple and highly reversible optical switch in a pixel is obtained, driven by electrowetting. These microscale pixels are arranged as demonstrated in Figure 1 to achieve the information display.

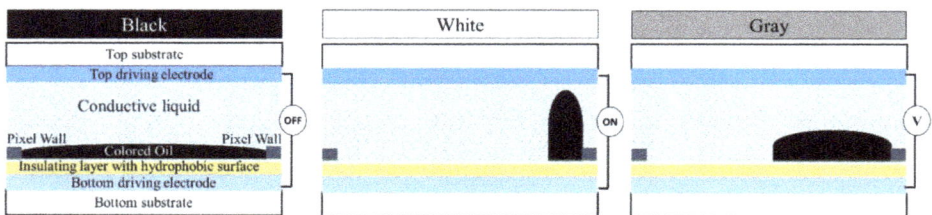

Figure 3. Device structure and working principle of the electrowetting display.

The performance of a EWD is highly dependent on the device's geometry and the properties of the properties of its constructive materials of the device, including an insulating layer, a hydrophobic layer, oil, and a conductive liquid. Based on the classical theory of electrowetting, the reduction of the contact angle is induced by the electrostatic force when a voltage is applied between the conductive fluid and electrode underneath [17]. When the applied voltage changes from 0 to V, an obvious force balance is rebuilt to induce the contact angle θ to change from θ_0 to θ_V. The electrostatic force is balanced by the surface energy. The electrostatic force is calculated by $CV^2/2$, where C is the capacitance between the electrode and the conductive liquid. As the electrode is covered by a hydrophobic insulating layer with a thickness of d, the capacitance is calculated as $C = \varepsilon_0\varepsilon_r/d$ (ε_0 and ε_r are the dielectric constants of air and the insulating material, respectively). Therefore, the obtained contact angle at voltage V (θ_V) is calculated by the Young–Lippmann equation [2,17]:

$$cos\theta_V = cos\theta_0 + \frac{\varepsilon_0.\varepsilon_r}{2\gamma_{wo}d}V^2, \tag{2}$$

where r_{wo} is the interfacial tension between conductive water phase and insulating oil phase. This equation has been successfully employed by many investigators in correlating experimental results with theory for a significant change in contact angles.

With the effort of scientists from both research institutions and companies, EWD has shown its potential for high-quality information displays. It has presented several advantages: (1) quick response with a switching speed <20 ms (>50 fps) for video display [18]; (2) good optical performance with >50% white state reflectance and full color range [2,8], and (3) the possibility of flexible displays due to the fully flexible fluidic display materials.

4. Interferometric Modulator Display

The interferometric modulator display (IMOD) is based on micro-electromechanical systems (MEMS) technology developed by the Iridigm Display Corporation (later merged into Qualcomm), which is also known as the Mirasol display [19,20]. The key of an IMOD is the tunable optical cavity. As shown in Figure 4, each optical cavity is composed of a self-supporting deformable reflective membrane on the bottom electrode and a thin-film stack residing on a transparent top substrate. Each stacks acts as one mirror of an optically resonant cavity. The deformable membrane is used to adjust the geometry of the cavity, which serves as the displaying pixel. When ambient light hits the structure, it is reflected off both the top of the thin-film stack (L_1) and the reflective membrane (L_2). Depending on the height of the cavity, reflected light of a certain wavelength will be slightly out of phase when reflecting off the membrane and the thin film. Therefore, some wavelengths will either constructively or destructively interfere depending on the phase difference. Human eyes will receive one color of a certain wavelength (e.g., red), which will be amplified by constructive interference with respect to others. However, the destructive interference results in a dark state (black). When a voltage is applied, the flexible membrane will be attracted up to tune the gap height driven by the electromechanical force. In this way, the displayed color (light) is selectively manipulated by the applied driving voltage.

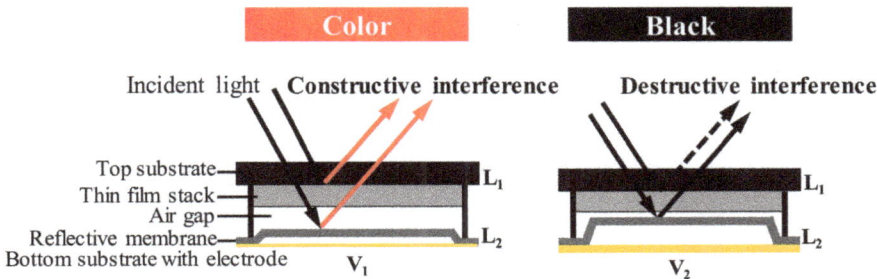

Figure 4. Device structure and working principle of the interferometric modulator display (IMOD).

An IMOD pixel is an optically resonant cavity similar to a Fabry–Pérot interferometer (etalon). When ambient light enters this cavity and reflects off the thin-film mirror, it interferes with itself, generating a resonant color determined by the height of the cavity. Therefore, the requirement for a Fabry–Pérot etalon is applied for the IMOD, for which,

$$h = m\left(\frac{\lambda}{2}\right), \tag{3}$$

where h is the height of the cavity, m is an integer, and λ is the wavelength of the light inside the cavity. In this way, the light reflection is controlled by the fluidic medium (the air cavity) to achieve the optofluidic display. A practical IMOD is composed of thousands of pixels with primary colors of red (R),

Micromachines **2018**, *9*, 159

green (G), and blue (B). A single RGB pixel is built from one or more subpixels. A monochromatic array of subpixels represents different levels for each color, and for pixels [19,20]. Therefore, in IMOD, by tuning the fluidic airgap through the applied voltage, not only is the displayed color selected from the ambient light (white), but also the brightness is determined depending on the uniformity of the constructed devices [3].

Depending on the electric field applied between the substrate and the thin film, the film can be positioned in one state that can remain until the next refresh [3,21]. Thus, a bi-stable reflective display is achieved. Flexible displays are also feasible with this technology by employing a flexible polyethylene naphthalate (PEN) film as the bottom substrate and suspended upper substrate [22]. However, due to the low production yield caused by sophisticated fabrication and a low display speed (~40 fps), IMOD was outside the scope of mainstream technology after a short period of "flash."

5. Conclusions and Outlook

Controlling the optical performance of displays via microfluidic behavior is one of the key areas of optofluidics. In this review, we selectively review three popular reflective display mechanisms using micro-confined fluidic mediums as displaying materials, including electrophoretic display (EPD), electrowetting display (EWD), and interferometric modulator display (IMOD). The driving forces are electrophoretic, electrowetting, and electro-mechanical forces in EPD, EWD, and IMOD, respectively. The microfluid-based displaying mediums are the nanoparticle dispersion, oil-water bifluidic system, and air cavity in EPD, EWD, and IMOD. The optical performance of these three display technologies is determined by the fluidic behavior confined in the pixels of each display device.

Nowadays, displays are everywhere in our life as key components of mobile phones, computers, televisions, etc. Reflective displays make use of ambient light, thus having advantages of comfortable readability in bright and outdoor environments, and energy savings according to bi-stable display and no backlight. In the future, with better control of fluidic properties (viscosity, surface tension, refractive index, conductivity, and dielectric constant) and the development of device design and fabrication, these optofluid-based reflective displays could achieve faster responses and higher optical contrast. New optofluidic devices and high-quality flexible displays can also be expected according to the fully flexible display materials based on fluids.

Acknowledgments: We appreciate the financial support from the National Key Research & Development Program of China (2016YFB0401502), the National Natural Science Foundation of China (61574065, 51561135014, U1501244), the Science and Technology Planning Project of Guangdong Province (2016B090906004), the Special Fund Project of Science and Technology Application in Guangdong (2017B020240002), and the Innovation Team of Zhongshan City (No. 170615151170710). This work has also been partially supported by PCSIRT Project No. IRT_17R40, the National 111 Project, the MOE International Laboratory for Optical Information Technologies, and the Cultivation Project of National Engineering Technology Center of Optofluidic Materials and Devices (2017B090903008).

Author Contributions: Mingliang Jin and Lingling Shui discussed and wrote the manuscript together. Shitao Shen, Zichuan Yi, and Guofu Zhou discussed and corrected the display technologies of EWD, EPD, and IMOD, respectively. Shitao Shen also helped with drawing the figures.

Conflicts of Interest: The authors declare no conflict of interest.

References

1. Chen, Y.; Au, J.; Kazlas, P.; Ritenour, A.; Gates, H.; McCreary, M. Flexible active-matrix electronic ink display. *Nature* **2003**, *423*, 136. [CrossRef] [PubMed]
2. Hayes, R.A.; Feenstra, B.J. Video-speed electronic paper based on electrowetting. *Nature* **2003**, *425*, 383–385. [CrossRef] [PubMed]
3. Liao, C.-D.; Tsai, J.-C. The evolution of mems displays. *IEEE Trans. Ind. Electron.* **2009**, *56*, 1057–1065. [CrossRef]
4. Heikenfeld, J.; Drzaic, P.; Yeo, J.S.; Koch, T. A critical review of the present and future prospects for electronic paper. *J. Soc. Inf. Disp.* **2011**, *19*, 129–156. [CrossRef]

5. Ota, I.; Ohnishi, J.; Yoshiyam, M. Electrophoretic image display (EPID) panel. *Proc. IEEE* **1973**, *61*, 832–836. [CrossRef]

6. Wang, L.; Yi, Z.; Jin, M.; Shui, L.; Zhou, G. Improvement of video playback performance of electrophoretic displays by optimized waveforms with shortened refresh time. *Displays* **2017**, *49*, 95–100. [CrossRef]

7. Comiskey, B.; Albert, J.D.; Yoshizawa, H.; Jacobson, J. An electrophoretic ink for all-printed reflective electronic displays. *Nature* **1998**, *394*, 253–255. [CrossRef]

8. Heikenfeld, J.; Zhou, K.; Kreit, E.; Raj, B.; Yang, S.; Sun, B.; Milarcik, A.; Schwartz, R. Electrofluidic displays using Young-Laplace transposition of brilliant pigment dispersions. *Nat. Photonics* **2009**, *3*, 292–296. [CrossRef]

9. Rogers, J.A.; Bao, Z.; Baldwin, K.; Dodabalapur, A.; Crone, B.; Raju, V.R.; Kuck, V.; Katz, H.; Amundson, K.; Ewing, J.; et al. Paper-like electronic displays: Large-area rubber-stamped plastic sheets of electronics and microencapsulated electrophoretic inks. *Proc. Natl. Acad. Sci. USA* **2001**, *98*, 4835–4840. [CrossRef] [PubMed]

10. Lu, C.M.; Wey, C.L. A controller design for color active-matrix displays using electrophoretic inks and color filters. *J. Disp. Technol.* **2011**, *7*, 482–489. [CrossRef]

11. Henzen, A. Development of e-paper color display technologies. *SID Symp. Dig. Tech. Pap.* **2009**, *40*, 28–30. [CrossRef]

12. Semenov, I.; Raafatnia, S.; Sega, M.; Lobaskin, V.; Holm, C.; Kremer, F. Electrophoretic mobility and charge inversion of a colloidal particle studied by single-colloid electrophoresis and molecular dynamics simulations. *Phys. Rev. E* **2013**, *87*, 022302. [CrossRef] [PubMed]

13. Beni, G.; Tenan, M.A. Dynamics of electrowetting displays. *J. Appl. Phys.* **1981**, *52*, 6011–6015. [CrossRef]

14. Jackel, J.L.; Hackwood, S.; Veselka, J.J.; Beni, G. Electrowetting switch for multimode optical fibers. *Appl. Opt.* **1983**, *22*, 1765–1770. [CrossRef] [PubMed]

15. Jackel, J.L.; Hackwood, S.; Beni, G. Electrowetting optical switch. *Appl. Phys. Lett.* **1982**, *40*, 4–6. [CrossRef]

16. Beni, G.; Hackwood, S. Electrowetting displays. *Appl. Phys. Lett.* **1981**, *38*, 207–209. [CrossRef]

17. Mugele, F.; Baret, J.C. Electrowetting: From basics to applications. *J. Phys. Condens. Matter* **2005**, *17*, R705–R774. [CrossRef]

18. Smith, N.R.; Hou, L.L.; Zhang, J.L.; Heikenfeld, J. Fabrication and demonstration of electrowetting liquid lens arrays. *J. Disp. Technol.* **2009**, *5*, 411–413. [CrossRef]

19. Miles, M.W.; Larson, E.; Chui, C.; Kothari, M.; Gally, B.; Batey, J. Digital paper™ for reflective displays. *J. SID.* **2003**, *11*, 209–215.

20. Sampsell, J.B. Mems-based display technology drives next-generation FPDs for mobile applications. *Inf. Disp.* **2006**, *22*, 24–28.

21. Miles, M.W. Mems-based interferometric modulator for display applications. In *Micromachined Devices and Components V*; French, P.J., Peeters, E., Eds.; International Society for Optical Engineering: Bellingham, WA, USA, 1999; Volume 3876, pp. 20–28.

22. Taii, Y.; Higo, A.; Fujita, H.; Toshiyoshi, H. Electrostatically controlled transparent display pixels by pen-film mems. In Proceedings of the IEEE/LEOS International Conference on Optical MEMS and Their Applications Conference, Oulu, Finland, 1–4 August 2005; pp. 13–14.

MDPI

St. Alban-Anlage 66

4052 Basel

Switzerland

Tel. +41 61 683 77 34

Fax +41 61 302 89 18

www.mdpi.com

Micromachines Editorial Office

E-mail: micromachines@mdpi.com

www.mdpi.com/journal/micromachines